Advances in Asian Human-Environmental Research

The *Advances in Asian Human-Environmental Research* series aims at fostering the discussion on the complex relationships between physical landscapes, natural resources, and their modification by human land use in various environments of Asia. It is widely acknowledged that human-environmental interactions become increasingly important in Area Studies and development research, taking into account regional differences as well as bio-physical, socio-economic and cultural particularities. The book series seeks to explore theoretic and conceptual reflection on dynamic human-environment systems applying advanced methodology and innovative research perspectives. The main themes of the series cover urban and rural landscapes in Asia. Examples include topics such as land and forest degradation, vulnerability and mitigation strategies, natural hazards and risk management concepts, environmental change, impact studies and consequences for local communities. The relevant themes of the series are mainly focused on geographical research perspectives of Area Studies, however there is scope for interdisciplinary contributions from various spheres within the natural sciences, social sciences and humanities. Key themes: Human-Environment Interaction - Asian regional studies - Asian geography - Impact studies - Landscape - Society - Land use and land cover change - Natural resources. *Submit a proposal*: Proposals for the series will be considered by the Series Editor and International Editorial Board. An initial author/ editor questionnaire and instructions for authors can be obtained from the Publisher, Dr. Robert K. Doe (robert.doe@springer.com).

Seema Rani

Climate, Land-Use Change and Hydrology of the Beas River Basin, Western Himalayas

 Springer

Seema Rani (iD)
Department of Geography, Institute of Science
Banaras Hindu University
Varanasi, Uttar Pradesh, India

ISSN 1879-7180   ISSN 1879-7199   (electronic)
Advances in Asian Human-Environmental Research
ISBN 978-3-031-29524-9   ISBN 978-3-031-29525-6   (eBook)
https://doi.org/10.1007/978-3-031-29525-6

Cover image: Nomads near Nanga Parbat, 1995. Copyright © Marcus Nüsser (used with permission)

This Springer imprint is published by the registered company Springer Nature Switzerland AG
The registered company address is: Gewerbestrasse 11, 6330 Cham, Switzerland

To my mother

Preface

The word "Himalaya" is derived from the Sanskrit words Hima, which means "snow," and Alaya, which means "abode." It is the world's tallest and youngest mountain range, located in the High Mountain Asia (HMA) region. It stretches around 2500 km from Nanga Parbat (8126 m) in Gilgit-Baltistan (Pakistan) to Namcha Barwa in Assam (India), with an average width of roughly 240 km. It covers an area of around 0.5 million km^2 in India, accounting for approximately 16.2% of the country's area. The Indian Himalayan Region (IHR) is divided into three parts: Himadri (Greater Himalaya), Himachal (Lesser Himalaya), and the Shiwaliks (Outer Himalaya). The IHR covers nine states (Himachal Pradesh, Uttarakhand (U.K.), Sikkim, Nagaland, Arunachal Pradesh, Mizoram, Manipur, Meghalaya, Tripura, hilly sections of two states (West Bengal and Assam), and two union territories (Ladakh and Jammu & Kashmir) of India. Aside from three folding axes or longitudinal sub-divisions, the IHR may be split into three areas based on regional characteristics, namely western Himalaya (Kashmir and Himachal Himalaya), central Himalaya (U.K. Himalaya), and eastern Himalaya (Darjeeling-Kalimpong-Sikkim-Arunachal Pradesh and Assam Himalaya and Purvanchal in Nagaland, Manipur, Mizoram, Tripura, and Meghalaya states). The IHR has diverse climatic and physiological conditions that support many forms of biodiversity. Many major rivers (such as the Ganga, Indus, and Brahmaputra) flow from this region, giving water to a considerable portion of the Indian subcontinent and directly or indirectly impacting the life of a vast population in the Himalayan and great plain regions for ages. This book is based on the climate, land-use change, and hydrology of the Beas River, a tributary of the Indus River, located in the states of Himachal Pradesh and the Punjab of India (Wester Indian Himalaya). It originates at the Beas Kund near the Rohtang Pass (Pir Panjal ranges) in the Kullu district of the state of Himachal Pradesh at an altitude of 4085 m above the mean sea level (AMSL). Various environmental, demographic, economic, and social systems also characterize the basin. The IHR's resources including this river support the lives of about 50 million people.

Concerns about climate change and other anthropogenic activities in the basin have been a topic of discussion. The interaction of climate, land cover changes, and water resources is becoming increasingly important in science and is considered critical to the general sustainability of the ecosystem in the twenty-first century. The scientific community is warning world leaders of the dangers of climate change and its inevitable effects on the physical and cultural environment. The Intergovernmental Panel on Climate Change (IPCC) recently stated in its recent special report on the Ocean and Cryosphere in a Changing Climate (SROCC) (2019) that reduced mass loss of the cryosphere by ice sheets/glaciers and reduced snow cover have implications for resource and aesthetic/cultural aspects of various parts of the IHR. It is causing severe weather occurrences in the IHR (warm winter and frequent cloud bursts). As a result, there is a need to understand how changes in climate and land cover impact the hydrology of the basin, which may substantially lead to better water resource planning and management. However, measuring these dynamic interactions in both space and time is difficult. Studies have been conducted across the world to examine the effects of climate and land cover changes on hydrology at various geographical scales. Nonetheless, because of the complicated climatic and land cover condition in the Himalayas, the findings cannot be abstracted from the basin size. This region has seen various physical changes over the years, including natural disasters (landslides, flash floods caused by cloud bursts), which have impacted the lives of millions of people. According to experts at the International Centre for Integrated Mountain Development (ICIMOD), Nepal, this Himalayan region is warming faster than the global average. Similar warming has also been noted in the Beas River basin. Climate change will influence the livelihoods of millions of people in the region. Therefore, for long-term planning and adaptation, a study on the effects and response of this mountainous region to climate and land use changes is essential.

In 2008, the Government of India (GoI) announced the National Action Plan on Climate Change (NAPCC) to promote regional sustainability while tackling climate change challenges. Among the NAPCC's missions, the National Mission for Sustaining the Himalayan Ecosystem (NMSHE) is one of the most significant, aiming to contribute to the country's sustainable development by improving the understanding of climate change, its expected implications, and adaptation activities for the Himalayas. In 2018, based on the recommendations of five thematic working group reports, the National Institute for Transformation of India (NITI) Aayog established the Himalayan State Regional Council for Sustainable Development to re-evaluate the implementation of the identified action points (inventory and revival of springs in the Himalayas for water security, sustainable tourism in the IHR, shifting cultivation: toward transformational approach, strengthening). To build plans under the given topics, large amounts of data are required, which is also a big worry owing to the region's harsh terrain and extreme weather conditions.

Climate and land cover changes have an impact on the hydrological conditions in the Beas basin. Rising air temperature in the Beas basin is causing glacial melt, snow cover area loss, and seasonal precipitation variance, all of which affect water availability in the basin. It has also been noticed that, while climate and land cover change occur everywhere, their consequences and amplitude vary from local to regional levels. Thus, this book is an attempt to discuss the climate, land use/land cover, and hydrology of the Beas basin and their relationships. The book would also emphasize on future flow regime conditions under different climate and land cover change scenarios.

In an integrated book titled *Climate, Land-Use Change and Hydrology of the Beas River Basin, Western Himalayas*, an attempt is made to grasp the existing data. This book aims to synthesize existing knowledge on the Himalayan basin dynamics and their consequences for physical systems. The book is primarily intended for academic academics, scientists, planners, and authorities involved in climate science. This book would be useful for people working on the physical components of the IHR in general and the basin in particular. This book offers a thorough and current review of different elements of climate and land cover changes and their effects on water, which will serve as a vital reference source based on accessible literature. I also hope that future scholars will expand on and contribute to the results presented in this book by employing up-to-date high-quality data and rapidly expanding geospatial technology.

Varanasi, Uttar Pradesh, India   Seema Rani
January 2023

Acknowledgments

Supervisor

This book is a continuation of my Ph.D. work, so I would like to express my heartfelt gratitude to my supervisor, Dr. S. Sreekesh (Centre for the Study of Regional Development, Jawaharlal Nehru University, New Delhi, India), who greatly supported my Ph.D. I also would like to express my gratitude to Jawaharlal Nehru University in New Delhi, India, where some of the research for the book was initiated.

Financial Support

I also express my sincere thanks to the Banaras Hindu University, Varanasi, Uttar Pradesh (India), for providing a seed grant (No. R/Dev/D/IoE/Equipment/Seed Grant- II/2022-23/52078) under the Institute of Eminence (IoE) which helps in the fieldwork of the book.

Expert Guidance

Dr. Suraj Mal, Department of Geography, Shaheed Bhagat Singh College, University of Delhi, New Delhi, India. Presently, Chair, International Geographical Union (IGU) Commission on Biogeography and Biodiversity.

Students

Ms. Jyotsna Singh, Ph.D. scholar, Department of Geography, Institute of Science, Banaras Hindu University, Varanasi, Uttar Pradesh, India. She assisted in the climate data of Chap. 3.

Mr. Amol Kumar Garg, M.Sc. student, Department of Geography, Institute of Science, Banaras Hindu University, Varanasi, Uttar Pradesh, India. He assisted in the basin morphometric analysis of Chap. 2.

Mr. Purusottam Tiwari, Ph.D. scholar, Department of Geography, Institute of Science, Banaras Hindu University, Varanasi, Uttar Pradesh, India. He assisted in the flood section of Chap. 4.

Data Provider

I am grateful to the United States Geological Survey (USGS) Earth Resources Observation and Science (EROS) Center for providing the Landsat data. I am also thankful to the Department of Space, the Indian Space Research Organization, and the National Remote Sensing Centre for providing Cartosat elevation data for the present study. I am also grateful to European Centre for Medium-Range Weather Forecasts (ECMWF) team and the MODIS science team for providing different data in the public domain. I am grateful to the Survey of India (SoI) and the Soil and Land Use Survey of India (SLUSI) for providing toposheets and soil data of the study area.

Series Editor

I am grateful to the springer series editor Prof. Marcus Nüsser, Heidelberg University, Germany for his thoughtful and thorough review that improves the quality of the book.

Springer Team

I would like to express my gratitude to Dr. Robert K. Doe, Executive Editor, and springer team who helped me with this book.

Contents

Abbreviations

AGCM	Atmospheric General Circulation Model
AMRUT	Atal Mission on Rejuvenation and Urban Transformation
AMSL	Above Mean Sea Level
ARS	Agricultural Research Service
AVHRR	Advanced Very High-Resolution Radiometer
BBMB	Bhakra Beas Management Board
BMB	Bhakra Management Board
BUW	Barren/Unculturable/Wasteland
C3S	Copernicus Climate Change Service
CASC2D	CASCade of planes, 2-Dimensional
CBH	Cloud Base Height
CC	Cloud Cover
CCAP	Climate Change Action Program
CCM	Constant Channel Maintenance
CDM	Clean Development Mechanism
CH	Central Himalayas
DAAC	Distributed Active Archive Center
DEM	Digital Elevation Models
DHSVM	Distributed Hydrology-Soil-Vegetation Model
DOS	Dark Object Subtraction
DWSM	Dynamic Watershed Simulation Model
ECMWF	European Centre For Medium-Range Weather Forecasts
EDW	Elevation-Dependent Warming
EEFP	Energy Efficiency Financing Platform
NSE	Nash-Sutcliffe Efficiency
ESIP	Ecosystems Services Improvement Project
ET	Evapotranspiration
ETM	Enhanced Thematic Mapper Plus
FAO	Food and Agricultural Organization
FEEED	Framework for Energy Efficient Economic Development
GBPIHED	G.B. Pant Institute of Himalayan Environment and Development

GHG	Greenhouse Gas
GHS	Global Human Settlement
GIM	National Mission for Green India
GIS	Geographic Information System
GoHP	Government of Himachal Pradesh
GPS	Global Positioning System
GPW	Gridded Population of the World
G-SHE	Governance for Sustaining the Himalayan Ecosystem
HADP	Hill Area Development Programme
HBV	Hydrologiska Byråns Vattenbalansavdelning Model
HCC	High Cloud Cover
HDR	Human Development Report
HKH	Hindu-Kush Himalaya
HMS	Hydrologic Modelling System
HP	Himachal Pradesh
HPSCCAP	Himachal Pradesh State Climate Change Action Plan
HRU	Hydrological Response Units
ICFRE	Council of Forestry Research and Education
ICIMOD	International Centre for Integrated Mountain Development
IH	Indian Himalayas
IHCAP	Indian Himalayas Climate Adaptation Programme
IHR	Indian Himalayan Region
IKI	International Climate Initiative
IMD	India Meteorological Department
INCCA	Indian Network for Climate Change Assessment
IPCC	Intergovernmental Panel on Climate Change
IRS	Indian Remote Sensing
ISMR	Indian Summer Monsoon Rainfall
ISRO	Indian Space Research Organization
JJAS	June-July-August-September
KP	Kyoto Protocol
LCC	Low Cloud Cover
LH-OAT	Latin Hypercube One-Factor-At-a-Time
LULC	Land Use/Land Cover
MCC	Middle Cloud Cover
MNRE	Ministries of New and Renewable Energy
MODIS	Moderate Resolution Imaging Spectroradiometer
MoEFCC	Ministry of Environment, Forest, and Climate Change
MONERIS	Modelling Nutrient Emissions in River Systems
MRI-JMA	Meteorological Research Institute-Japan Meteorological Agency
MSS	Multi-Spectral Scanner
MTEE	Market Transformation for Energy Efficiency
NAFCC	National Adaptation Fund for Climate Change
NAPCC	National Action Plan for Climate Change
NCAP	National Carbonaceous Aerosols Programme

NCEPP	National Council for Environmental Policy And Planning
NDBI	Normalized Difference Built-Up Index
NDCs	Nationally Determined Contributions
NDSI	Normalized Difference Snow Index
NDVI	Normalized Difference Vegetation Index
NEP	National Environment Policy
NFP	National Forest Policy
NITI	National Institution for Transforming India
NMEEE	National Mission for Enhanced Energy Efficiency
NMSA	National Mission for Sustainable Agriculture
NMSH	National Mission on Sustainable Habitat
NMSHE	National Mission for Sustaining the Himalayan Ecosystem
NMSKCC	National Mission on Strategic Knowledge for Climate Change
NOAA	National Oceanic and Atmospheric Administration
NRSC	National Remote Sensing Centre
NSIDC	National Snow and Ice Data Center
NSM	National Solar Mission
NWM	National Water Mission
OLI TIRS	Operational Land Imager and Thermal Infrared Sensor
OLS	Ordinary Least Square Regression Test
OM	Organic Matter
PAT	Perform Archive and Trade
PET	Potential Evapotranspiration
PMCCC	Prime Minister Council on Climate Change
PRECIS	Providing Regional Climates For Impact Studies
PRMS	Precipitation-Run-Off Modelling System
PWV	Precipitable Water Vapor
RCMs	Regional Climate Models
RCP	Representative Concentration Pathways
SAPCC	State Action Plan on Climate Change
SCA	Snow Cover Area
SCS	Soil Conservation Service
SDC	Swiss Agency for Development and Cooperation
SLADRC	Sustainable Livelihoods of Agriculture-Dependent Rural Communities
SLUSI	Soil and Land Use of Survey India
SMOD	Settlement Model Grid
SOI	Survey of India
SRES	Special Report on Emissions Scenarios
SRM	Snowmelt-Run-Off Model
SUFI	Sequential Uncertainty Fitting
SWAT	Soil and Water Assessment Tool
SWAT-CUP	Calibration and Uncertainty Programs
SWIM	Soil and Water Integrated Model
TCC	Total Cloud Cover

TM	Thematic Mapper
TOA	Top-of-Atmosphere
TP	Tibetan Plateau
UBDC	Upper Bari Doab Canal
UNEP	United Nations Environment Programme
UNFCCC	United Nations Framework Convention On Climate Change
UNSC	United Nations Security Council
USDA	United States Department of Agriculture
USLE	Universal Soil Loss Equation
UTM	Universal Transverse Mercator
WASIM	Water Balance Simulation Model
WASMOD	Water and Snow Balance Modeling System
WDP	Watershed Development Programmes
WGS	World Geodetic System
WIH	Western Indian Himalaya
WMO	World Meteorological Organization
WRI	Water Ratio Index
WRIS	Water Resources Information System

Chapter 1
Introduction

Abstract The High Mountain Asia (HMA) is home to the tallest and youngest mountain range (e.g., Himalayas) of the globe. The term "Himalaya" is derived from the Sanskrit words Hima (snow) and Alaya (abode). With an average breadth of around 240 km, it spans over 2500 km from Nanga Parbat (8126 m) in Gilgit-Baltistan (Pakistan) to Namcha Barwa in Assam (India). It occupies an area of around 0.5 million km² in India or 16.2% of the total land area of the nation. Many types of biodiversity are supported by the varied meteorological and physiological conditions found in the Indian Himalayan Region (IHR). This area is the source of several important rivers, including the Ganga, Indus, and Brahmaputra, which provide water to a substantial portion of the Indian subcontinent and have long had a direct or indirect influence on the lives of a sizable population in the Himalayan and great plain areas. Science is paying more and more attention to how climate, land cover changes, and water resources interact since it is thought that this is essential for the ecosystem's overall sustainability in the twenty-first century. Global leaders are being alerted to the risks posed by climate change and its unavoidable repercussions on the natural and cultural environment by the scientific community. There is a need to comprehend the growing interactions between climate, land use change, and hydrology in the Beas River basin. This would be helpful for planners and stakeholders who are working on the same theme.

Keywords High Mountain Asia · Himalayas · Indian Himalayan Region · Indus River · Beas River · Climate change · Hydrology · Land use/land cover

1.1 Background

The atmosphere and land processes are complex and need attention to overcome the ongoing challenges of global warming. Changes in the climate and land use/land cover (LULC) are a global problem for sustainability. The total land and ocean surface temperatures throughout the world have risen to around 0.89 °C since 1901 and

S. Rani, *Climate, Land-Use Change and Hydrology of the Beas River Basin,*
Western Himalayas, Advances in Asian Human-Environmental Research,
https://doi.org/10.1007/978-3-031-29525-6_1

0.72 °C over the past three decades, showing the fastest increase in surface temperature since the turn of the century (IPCC 2013). Long-term changes to global, regional, and local hydrology would have an impact on millions of people's ability to live comfortably. Likewise, changing precipitation and snow melting are also modifying hydrological systems in many regions (IPCC 2013; Pörtner et al. 2022). Climate change is also causing permafrost to warm in high-latitude and high-altitude regions (IPCC 2013; Pörtner et al. 2022). India is now dealing with the consequences of climate change. Surface air temperature in India has increased by 0.4 °C over the previous century, which will cause glaciers to melt and would eventually have a detrimental impact on the water supply (NATCOM 2012; Huss and Hock 2018).

Water demand is rising as a result of the expanding global population and the pressing need for water resource management, in addition to other developments. The expected increases in world population from 7.7 billion in 2019 to 8.5 billion in 2030 and 9.7 billion in 2050 would change the LULC (United Nations 2019). In Asia, where the population is estimated to expand from 4.5 billion in 2017 to 5.2 billion in 2050, additional population growth is anticipated (United Nations 2019). The worldwide water supplies needed for both the natural ecology and the livelihood of the population are, however, rapidly depleting. Inaccessibility to safe drinking water affects one in five people (World Water Council 2000). By 2050, 4 billion people, or more than half of the world's population, would reside in nations where more than 40% of renewable resources are used for human consumption. By 2025, 0.07 billion people in India won't have access to safe and affordable drinking water (World Water Council 2000). According to projections by the United Nations (2019), India's population would increase from 1.34 billion in 2017 to 1.66 billion in 2050, making it the world's biggest nation by 2027.

It is also projected that climate and LULC changes may exacerbate the water-deficit scenario by changing the hydrological cycle and affecting the availability of water resources. It is possible to see the impact of climatic variability on the parameters of the hydrological cycle not only on a global and national scale but also at the local level. In the Western Indian Himalaya (WIH), the Beas River basin is experiencing seasonal precipitation fluctuations, a reduction in snow cover area (SCA) (Rani 2014), and glacier melt due to rising air temperature (Kulkarni et al. 2005). The availability of water in the basin would be impacted by this. Water availability has an impact on practically every aspect of human existence, including agricultural production, energy consumption, municipal and industrial water supply, flora and fauna management, and general care of terrestrial and aquatic ecosystems. As a result, understanding the connection between climate, land cover, and hydrology is becoming increasingly important in the scientific community for effective water resource planning and management. It is also critical to the overall sustainability of the Earth's system.

1.2 Emerging Climate Change Scenario

1.2.1 Air Temperature

Studies around the globe show evidence of rising temperature (Ren et al. 2017; Mal et al. 2018, 2020; Krusell and Smith 2022; Rani and Mal 2022). Since the end of the nineteenth century, the average global surface temperature has risen by almost 0.6 °C (IPCC 2007), 0.89 °C (1901–2012), and over 0.72 °C (1951–2012) (IPCC 2013). Maximum and minimum land temperatures have increased globally since 1950. The 2000s were the warmest decade in the past three decades, which were warmer than any of the decades before them. Over Central and South Asia (1961–2000), there has been a statistically significant decline in the percentage of cold nights and days and an increase in the percentage of warm nights and days (Klein Tank et al. 2006). The earth system as a whole is warming, while there are localized variances. The 6th Assessment Report of the Intergovernmental Panel on Climate Change (IPCC) (IPCC 2021a) estimates that the global temperature may exceed 1.5 °C of warming in the next two decades, i.e., by 2040, resulting in higher heat waves, long summers, and shorter winter seasons. The 2011–2020 decade is anticipated to be hotter than any previous period in the preceding 1.25 lakh years, with intense rain occurrences intensifying by almost 7% for every 1 °C increase in temperature. Aside from that, both annual and monsoonal precipitations are anticipated to rise globally, with more inter-annual variability across Southeast Asia. According to Ali et al. (2019a, b), the upper Indus basin (UIB) region's overall mean temperature decreased by 0.137 °C in Gilgit while rising by 0.63 °C in Skardu during the years 1980 and 2006 and 1953 to 1979, respectively. The same study observed a decrease in the mean minimum temperature, while the mean maximum temperature showed non-significant changes during the summer at both locations. During the period 1955–2016 for the UIB, Hussain et al. (2021) reported a broad significant rising trend of 0.14 °C/decade for the maximum temperature but a significant falling trend of −0.08 °C/decade for the annual minimum temperature. Seasonally, winter and spring see more maximum temperature warming, whereas summer and autumn experience greater minimum temperature cooling. Results of seasonal maximum temperatures show rising tendencies in winter, spring, and autumn at rates of 0.38, 0.35, and 0.05 °C/decade, respectively, whereas summer temperatures are falling at a rate of −0.14 °C/decade. Additionally, seasonal minimum temperature findings for the whole UIB show increasing trends in winter and spring at rates of 0.09 and 0.08 °C/decade, respectively, with considerably declining trends in summer and autumn at rates of −0.21 and −0.22 °C/decade, respectively. Furthermore, the report of IPCC (2021a) suggests that the alarming rate of global warming is likely to have serious consequences around the world due to the rapid melting of glaciers and retreating snowlines, which can alter the water cycle and precipitation patterns, leading to increased flood and drought events and increased water scarcity.

Evidence of warming is also observed at the India level, though there are regional variations in the rate of warming (Jhajharia and Singh 2011; Ray et al. 2019; Mal et al. 2020). According to the India Meteorological Department's annual climate summary (IMD 2019), there has been a warming trend in the annual mean air temperature at a rate of 0.061 °C/decade, with notable warming in the maximum air temperature (0.1 °C/decade) and relatively lower warming in the minimum air temperature (0.022 °C/decade) during 1901–2019. In India, several areas have seen warming between 1941 and 2012 at rates of 0.044 °C/decade (average temperature), 0.051 °C/decade (maximum temperature), and 0.019 °C/decade (minimum temperature) (Rani and Sreekesh 2018; Saxena and Mathur 2019; Ray et al. 2019).

Studies have also demonstrated that the air temperature in the Himalayas fluctuates significantly; however, the intensity of the variation differs at different times based on the locations and seasons (Khattak et al. 2011; Dimri and Dash 2012; Rani 2014; Rani and Mal 2022). According to Schickhoff et al. (2016), warming in the majority of the Himalayan areas occurred faster than the global mean trend (0.85 °C) between 1880 and 2012, with winter temperatures generally trending higher than those of other seasons. Between 1901 and 2014, the annual mean temperature increased by more than 0.3 °C/decade in the Tibetan Plateau (TP) region and by roughly 0.2 °C/decade across the eastern side of the Hindu-Kush Himalaya (HKH) range (Ren et al. 2017). Elevation-dependent warming (EDW), which is seen at higher altitudes of the TP (>4000 m), is responsible for the region's even more warming (Liu et al. 2009; Krishnan et al. 2019a). Based on reconstructions of the Indian Himalaya and Karakoram area of the Himalayas using tree rings, several studies discovered a substantial increasing/decreasing pattern in summer temperature and rainfall during the previous three to four centuries (Borgaonkar et al. 1994, 1996; Pant et al. 1998; Yadav et al. 1999; Hughes 2001). Under representative concentration pathways (RCPs), a statistically significant warming projection (0.3–0.9 °C/decade) is found for the Indian Himalayan Region (IHR) in all seasons (Dimri et al. 2019). Ren et al. (2017) discovered both a rise and fall in temperature trends throughout the HKH area between 1901 and 1970. In the IHR, a substantial warming trend (at sig level of 0.10) was revealed between 1980 and 2020, with an average warming rate of 0.36 °C/decade ranging from 0.25 to 0.53 °C/decade, showing regional variability (Rani et al. 2022). The warming is also notable for its impact on winter's minimum and maximum temperatures (0.17 °C/decade). Additionally, a warming trend was seen in the WIH between 1975 and 2006 for the mean maximum temperature (1.1–2.5 °C) in the winter months (December–February) (Dimri and Dash 2012). According to Dimri et al. (2019), the minimum and maximum air temperatures in the WIH have significantly warmed. Winter months over the WIH between 1980 and 2020 saw the highest rate of warming (up to 0.42 °C/decade) (Rani et al. 2022).

In the summer, the annual temperature increase over northern India's Sichuan Basin and Karakoram range is less than 0.10 °C per decade (Forsythe et al. 2017). The observed warming signal was attributed by You et al. (2017) to the rising human activities that are raising greenhouse gas concentrations. The Satluj River Basin in the WIH saw the fastest rate of warming at the highest altitude station (Kaza) under

RCP 8.5 (0.84 °C/decade) (Gupta et al. 2020). The pace of warming in the Himalayas is higher than the world average, demonstrating the area's susceptibility to climate change and its effects (Shrestha et al. 2012; Pepin et al. 2015). However, there are uncertainties in both observational research and climate change estimates in the HKH that are based on modeling (Mayewski et al. 2020). Furthermore, earlier snowmelt and a shorter winter might affect river flow regimes due to increased maximum and minimum air temperatures throughout the pre-monsoon season. Most of the mentioned studies on air temperature changes showed that the region's maximum and minimum temperatures were warming (particularly in winter). This would most likely result in a change from snow to rain in the form of precipitation. In many nations, it seems that more total precipitation falls as rain than snow. Long-term reductions in snowfall and snow depth would come from this which will influence the hydrology.

1.2.2   Precipitation

In addition to air temperature, evidence of a worldwide rise in annual and seasonal precipitation was observed (Westra et al. 2013; IPCC 2013; Bevacqua et al. 2022), while there are regional differences in the trend's intensity. For areas in the UIB at high elevation (>1300 m), Latif et al. (2018) reported declines in precipitation to look robust and more substantial. Krakauer et al. (2019) discovered a 15% increase in mean precipitation between 1891 and 2016 in the Indus basin. No discernible precipitation pattern was seen for the more recent history, dating back to 1958 or 1979. According to Hussain et al. (2021), annual precipitation increases at a rate of 2.74 mm/decade, whereas rates for the UIB's winter, summer, and autumn are 1.18, 2.06, and 0.62 mm/decade, respectively. In the UIB, Liaqat et al. (2022) showed a little increase in annual precipitation between 1995 and 2017, but a large increase in winter precipitation.

India is likewise dealing with rising and falling trends in precipitation (Guhathakurta and Rajeevan 2008; Kumar et al. 2010; Rai et al. 2010; Pal and Al-Tabbaa 2011; Mal et al. 2022). According to IMD (2016), the west coast and the core monsoon area of north-central India are experiencing a major decrease in the summer monsoon. Overall, although showing a decline in yearly and monsoon rainfall in some regions, the aforementioned studies did not identify any particular pattern of decreasing rainfall over all of India. Reduced monsoon rainfall would have an impact on the population's livelihood because India receives the majority of its rainfall during this season which is also the primary source of irrigation for agriculture.

Researchers examined the Western Himalayas' precipitation patterns (Guhathakurta and Rajeevan 2008; Shekar et al. 2010; Bhutiyani et al. 2010; Dimri and Dash 2012; Khattak et al. 2011; Mal et al. 2022). Although the Western Himalayas experienced average annual rainfall (1901–2013), there were periodic rainfall excesses in the winter and monsoon and deficits in the pre and post-monsoon (IMD 2016). Years from 1957 to 2007, Singh and Mal (2014) noted a decreasing tendency in the state of Uttarakhand's annual rainfall at high altitudes (located in the WIH). However, in low elevations, monsoon precipitation rose, whereas trends in winter rainfall were inconsistent. In the Himalayas, summer precipitation shows rather declining tendencies (Schickhoff et al. 2016). The "Karakoram Anomaly" is an increasing tendency of wintertime frozen precipitation that has been seen in several areas of the Karakoram Himalayas in recent years (Kapnick et al. 2014; Kääb et al. 2015; Krishnan et al. 2019). In the northwest Himalayan Karakoram area, the same tendency was also noted. But throughout the Central Himalayas (CH), precipitation showed a declining trend (Cannon et al. 2015; Madhura et al. 2015; Krishnan et al. 2019). In the northern Hindu Kush and central India, Zhan et al. (2017) noted a considerable shift in the intensity of light precipitation. According to RCP 8.5 scenarios, Gupta et al. (2020) discovered the Satluj River had the highest rate of precipitation loss (6.362 mm/year) at a low-altitude station (Kasol). But under RCP 2.6, the same analysis discovered a tendency toward more precipitation. Such changes under various RCPs in various elevation zones need to be examined with orographic processes in more depth. Immerzeel et al. (2015) found that high-altitude precipitation in this region is underestimated and that the large glaciers here can only be sustained if high-altitude accumulation is much higher than most commonly used gridded data products. Total cloud cover above the IHR has been decreasing significantly since 1980, with a mean rate of −0.0004%/decade ranging from −0.03 to 0.03%/decade (Rani et al. 2022). For the whole winter (Dec–Mar), the total clouds in the area significantly decreased (Rani et al. 2022). A detailed understanding of the cause-and-effect connection between rainfall and cloud cover across the area is required by employing updated spatial technologies.

When compared to other regions, Manali (located in the upper section of the Beas basin) showed high intra-seasonal fluctuation in rainfall (Rani 2014). The study found no clear pattern in the rainfall variation across Manali, Bhuntar, and Banjar; however by taking into account the average amount of rainfall across all stations, it can be stated that March has a significant falling trend in rainfall in the basin at a rate of 2.925 mm/year. As a consequence, a statistically significant downward trend in the amount of rainfall during the winter season (−5.52 mm/year) was found (Rani 2014). There is no seasonal or monthly pattern in the research area's average number of rainy days. It shows that the number of wet days both seasonally and monthly is stable and that rainfall in March is declining, suggesting that the intensity of the rain is declining. Most studies have revealed a general decline in precipitation in the Western Himalayas, which is alarming since it would mean less snow accumulation, which would have an adverse long-term effect on water availability.

1.2.3   Future Climate Scenario

With the aid of regional climate models (RCMs), efforts were also made to estimate the national climate change scenario based on the current climate conditions. Yadav et al. (2010) used the Providing Regional Climates for Impact Studies (PRECIS) and Meteorological Research Institute-Japan Meteorological Agency (MRI-JMA) and Atmospheric General Circulation Model (AGCM) to create high-resolution climate change scenarios for Northwest India's winter season. By the end of the twenty-first century, models conducted under high-emission scenarios indicate that precipitation and surface temperature will have significantly increased. According to Sarthi et al. (2012), the lower portion of India's western and eastern coasts would experience deficits and surpluses in rainfall during the summer monsoon depending on the emission scenarios A2, B1, and A1B. Additionally, Kumar et al. (2013) simulated potential climate change over South Asia using three incredibly high-resolution RCMs, with a focus on India. By the end of the twenty-first century, monsoon precipitation increased, with regional variability. By the end of the century, India will see average warming of 1.50 °C, down from 3.90 °C between 1970 and 1999 (Kumar et al. 2013). India's anticipated mean temperature is quite near to the 40 °C global average predicted by the IPCC. The projections showed a rise in precipitation over peninsular India and coastal regions but either no change or a drop farther inland, demonstrating substantial regional variation in precipitation trends. By the end of the twenty-first century, the A1B scenario projects that there will be widespread warming (−3.20 °C) and an overall increase (8.5%) in mean monsoon precipitation (Kumar et al. 2013). By the end of the twenty-first century, Kulkarni (2013) estimated that the HKH region will have warmed significantly. According to Forsythe et al. (2014), the UIB's mean temperature will rise year-round (annual mean +4.8 °C) (2071–2100). According to Nazeer et al. (2022), the UIB saw a significant rise in temperature (1.1–8.6 °C) throughout the course of the twenty-first century. In the Indo-Ganga-Brahmaputra basin, Wijngaard et al. (2017) discovered a rise in the amplitude of climatic means and extremes toward the end of the twenty-first century, with climatic extremes tending to become higher than climatic means. According to Shrestha et al. (2015), the Hindu Kush Himalayan region's rugged terrain will see an average temperature increase of 1–2 °C (and in some locations as much as 4–5 °C) by 2050. According to Ramzan et al. (2019), downscaled climate data suggest an increase in the mean maximum (0.3–2.3 °C) and minimum temperature (0.3–1.9 °C) compared to the baseline period of 1980–2010. The future (2075–2099) temperature is predicted to rise in every season across the UIB, with the biggest increase estimated to occur in September and October at about 8 °C, according to Baig et al. (2021) while the temperature will rise by 5 °C/year.

Additionally, according to Kulkarni (2013), summer monsoon precipitation would increase by 20–40% in the HKH region between 2071 and 2098 compared to the baseline (1961–1990). According to Forsythe et al. (2014), future precipitation in the UIB is expected to rise year-round (maximum seasonal mean change: +27%, annual mean change: +18%) with increased intensity in the wettest months (February, March, and April) (2071–2100). According to Nazeer et al. (2022), the amount of precipitation in the UIB is expected to rise significantly (12–32%) throughout the twenty-first century. According to Rajbhandari et al. (2015), the Indus River basin would see a rise of more than 40 °C by the 2080s. The study also found that precipitation was increasing across the UIB and decreasing over the lower Indus basin. Winter precipitation is expected to decrease, mostly across the southern section of the basin. Projections show that the UIB will warm more than the lower Indus and that winter will warm more than other seasons. Over the basin, a rise in the number of wet days was also simulated. At different geographical scales, climate forecasts under different scenarios indicated a decrease in winter precipitation and an increase in total precipitation. As per Shrestha et al. (2015), precipitation across the Hindu Kush Himalayan region will change by 5% on average and up to 25% by 2050. The monsoon is expected to become longer and more erratic. Extreme rainfall events are becoming less frequent, but more intense and are likely to keep increasing in intensity. According to Ramzan et al. (2019), at multiple meteorological locations and under different RCPs, an increased tendency for precipitation has been seen varying from 2% to 17% compared to the baseline period of 1980–2010. Autumn exhibits the greatest seasonal variation in both temperature and precipitation, followed by spring. According to Baig et al. (2021), future (2075–2099) precipitation will increase every month across UIB—specifically the increase in July–August period is very significant. Albeit the precipitation increase is highly variable on spatial scale, i.e., decreasing/increasing at same time in different regions. Climatic projections only indicate what climate conditions could be like in the near future. However, the forecasts are fraught with uncertainties. As a result, a greater understanding of current climate conditions is critical for analyzing their impact on available water supplies.

1.3   Land Use/Land Cover Changes

The Food and Agricultural Organization (FAO)[1] defines land use as the configurations, actions, and inputs that humans make in a certain form of land cover to produce, modify, or sustain it, such as a built-up region. The observed (bio) physical cover of the Earth's surface, such as water bodies, snow, grassland, desert, and so on, is referred to as land cover. It is a dynamic process that has developed over time as a result of interactions between people and their surroundings. Aside from climatic variability, the LULC pattern is an important component in determining a region's hydrological behavior. LULC changes simply refer to changes in the Earth's terrestrial surface caused by anthropogenic activities, where land-use refers to the use of land by humans for various purposes, and land-cover refers to the surface area on the ground covered by natural or artificial structures (such as water bodies, natural vegetation, crop cultivation, bare earth, buildings, dams, and so on) (Mondal and Zhang 2018). LULC changes in the Himalayas have been reported by several studies over time (Ullah et al. 2019; Fayaz et al. 2020; Mishra et al. 2020; Rani and Sreekesh 2022). The rising population has been modifying the land for centuries to meet their basic needs of food and shelter, but remarkable changes in LULC scenarios have occurred since the twentieth century due to unprecedented

---

[1] http://www.fao.org/docrep/003/x0596e/x0596e01e.htm

population growth, rapid industrialization, increasing urbanization, infrastructural development, and socioeconomic development occurring globally. These changing LULC scenarios, such as the expansion of settlement areas, agricultural land, grassland, and so on, at the expanse of forest areas or natural vegetation, affect water balance components such as evapotranspiration (ET), surface runoff, base flows, groundwater recharge, and so on, and as a result, the region's hydrological behavior is affected.

1.4 Impact of Climate and LULC Changes on Hydrology

According to Azam et al. (2021), the Indus, Ganges, and Brahmaputra River systems, which originate in Hindu Kush's glaciers and snowfields, supply water to 1 billion people. The Hindu Kush River basins have the greatest irrigated area and the highest built hydropower capacity (26,000 MW) (around 577,000 km^2). By changing hydrological processes like ET, accumulation and snowmelt processes, soil moisture storage, surface flow, water yield, and other processes, climate change and LULC changes affect the river basin's hydrological behavior, which in turn affects the availability of water in a region (Nepal and Shrestha 2015; Nepal 2016; Rani et al. 2019; Li et al. 2019a, b; Ali et al. 2019a; Bhatta et al. 2019; Bilal et al. 2021; Rani and Sreekesh 2021). Studies and reports suggest that the mountainous regions (such as the Rocky Mountains, the Andes, the Alps, and the Himalayas) experience more rapid warming in comparison to the downstream low-land areas because of EDW, which implies that some mechanisms (e.g., changes in cloud cover and soil moisture with altitude) cause more rapid warming with increasing altitude of the region (Singh et al. 2016; González-Zeas et al. 2019). The observed global warming is likely to have serious implications such as intensification of the water cycle, changes in precipitation patterns, melting of glaciers and ice sheets, alteration in SCA, rising mean sea level, etc., and all these changes may get aggravated with further warming (IPCC 2021b).

According to Shrestha et al. (2015), the Indus basin's annual water quantities won't change by 2050. However, the Indus basin glaciers will continue to lose a significant amount of mass. You et al. (2017) reviewed studies dealing with the issue of climate change in the HKH region. This review suggested that rapid warming causes solid-state water (i.e., snow, ice, glacier, and permafrost) to melt rapidly, which when coupled with higher precipitation or snowfall may lead to more frequently occurring events of flash floods, rock avalanches, landslides, and other disasters in the region (You et al. 2017; Krishnan et al. 2019). Other studies assessing the effect of climate change on the flow regime of Kalig and aki basins (Bajracharya et al. 2018) and Koshi basin (Nepal 2016), located in the HKH region, suggest that the significantly rising average annual temperature and average annual precipitation in the twenty-first century leads to an increase in the probable discharge and water yield, on the account of the rapid melting of snow and glacier. Thus, there seems to be no problem with water availability in these basins at least in this century, but the hydrological regime of the basins is likely to change under the influence of predicted climatic patterns, affecting water distribution (timing and quantity) in the downstream region. Although this climate variability may be initially beneficial in terms of the availability of water, its negative effects such as floods and glacial lake outburst floods (GLOFs) are unavoidable and require proper planning for water resource management in the region.

GLOFs are a major concern in the Himalayas in the last few decades (Nie et al. 2018; Schmidt et al. 2020; Compagno et al. 2022). Himalayan glacial lakes were explored by Worni et al. (2013). The study estimates that the Himalayan glaciers cover 17% of the mountain range and hold around 12,000 km^3 of freshwater. Additionally, they contain 251 glacial lakes larger than 0.01 km^2, and it was found that the properties of the lakes differed greatly depending on the location. Schmidt et al. (2020) estimated a total of 192 glacial lakes (5.93 ± 0.70 km^2) with an estimated water volume of about 61.11 ± 8.5 million m^3. It includes 127 proglacial (PG) and 9 supraglacial lakes and 56 lakes located on recent moraines (RM) in 2018 in Gya, Ladakh (Upper Indus basin). However, according to the change detection assessments, 22 glacial lakes perished (decreased by more than 90%) during 1969–2018, 4 disappeared during 1969–1993, 9 during 1993–2002, and 9 during 2002–2018. Nie et al. (2018) came to the conclusion that glacial lakes with rapid growth and a high likelihood of repeating their outbursts need more monitoring. According to Nie et al. (2021), the diverse glacier retreats that are changing streamflow patterns are increasing the frequency of GLOFs and raising the risk of flooding and water shortages associated with future climate change.

Moreover, Bajracharya et al. (2018) suggested that water balance components like snow melt, evapotranspiration, and water yield are likely to get more affected by climate change at higher altitudes of the upper and mid-sub-basin of the Kaligandaki basin as compared to basins at lower altitude. According to the RCP4.5 and RCP8.5 scenarios, Hassan et al. (2019) calculate streamflow for future climate change projections in the Northwestern UIB. Significant streamflow variations throughout the winter and spring seasons were predicted. Overall, the UIB should expect to experience more floods as a result of the anticipated rise in medium and high flow. By the end of the twenty-first century, it was discovered that the average yearly runoff was steadily rising. For the study period of January 2003 to December 2016, Hussain et al. (2020) assessed the spatial and temporal changes of terrestrial water storage in the UIB and discovered negative trends of −4.35–0.38 mm/year. The Indus-Tarbela inflows are more likely to rise than Kabul, Jhelum, and Chenab River inflows (Dahri et al. 2021). For all river gauges in the Indus basin, extreme climatic scenarios predict a significant rise in peak flow magnitudes and attainment 1 month early. A majority of studies found the changes in precipitation patterns to be heterogenous and more uncertain, while a homogenously increasing trend has been observed in the case of temperature conditions worldwide (Nepal and Shrestha 2015; Nepal 2016; Dimri et al. 2019; Rani and Sreekesh 2018).

According to Singh et al. (2016), the glacio-hydrological proxies (glaciers, glacier mass balance, and streamflow in downstream locations) for the IHR are more obvious indicators of current climate change than they were in the past. Frey et al.

(2012) created a glacier inventory of the Western Himalayas using satellite data. The report claims that more than 15,000 Himalayan glaciers feed enduring rivers like the Indus, Ganga, and Brahmaputra, which in turn sustain millions of people. The study found that the Western Himalayas had 11,400 glaciers larger than 0.02 km^2, spanning a total area of 9310 km^2, and that they have distinct patterns of mean glacier height and relative debris cover amounts that may be related to the local climate. Schmidt and Nüsser (2017) created an inventory of the glaciers and their alterations in the Trans-Himalaya in central and eastern Ladakh (Upper Indus basin). In general, they discovered that Ladakh's glaciers stand out for their high altitude (91% of them end above 5200 m) and relatively tiny size (79% are smaller than 0.75 km^2 and only 4% are more than 2 km^2). In the glaciated area of central Ladakh, there were more than 1800 glaciers covering 997 km^2 in 2002. The study found that the area of glaciers have significantly reduced. In Chandra basin, Western Indian Himalaya, Sahu and Gupta (2020) estimated that 395 glaciers are identified spanning 703.3 ± 20.4 km^2 area with minimum glacier size being >0.02 km^2. Fifty-nine of these glaciers, totaling 67.2 ± 1.9 km^2, have ice that is coated with debris. Estimated glacier area in 1971 was 639.4 ± 5.8 km^2, which fell to 620.2 ± 18.0 km^2 (−3.0 ± 3.0%) in 2002 and to 608.1 ± 10.3 km^2 (−4.9 ± 1.9%) in 2016. Using remote sensing data, Shukla et al. (2020) created an inventory of the Suru sub-basin in the Western Himalaya for the year 2017. According to inventory data, the sub-basin features 252 glaciers that span 11% of the basin and have an average slope of 25 ± 6° north. The results of a temporal study show that during the last 46 years, the glacier has shrunk by ~6 ± 0.02%, has been retreating at an average pace of 4.3 ± 1.02 m per year, and has increased its snow line altitude (SLA) by 22 ± 60 m. In the Western Himalaya's Chandra-Bhaga Basin (CB Basin), Vatsal et al. (2022) report a homogeneous, multidecadal inventory of glaciers using remote sensing data for the years 1993, 2000, 2010, and 2019. Two hundred fifty-one glaciers with an extent more than 0.5 km^2 have been found and manually mapped, while 6 glaciers in the basin underwent field surveys to reduce ambiguity. Two hundred seventeen of these 251 glaciers have clear ice, while 35 have glaciers covered with debris. In 1993, it was estimated that there were 996 km^2 of glaciers overall; in 2019, that number had dropped to 973 km^2. These glaciers provide water to vital rivers in India and are crucial for understanding how climate variability affects the Himalayan region. Half a billion people in the HKH region may be significantly impacted by the melting glaciers brought on by climate change (Stern 2007). In some regions, such as the Nanga Parbat region, Hunza-Karakoram, and Ladakh of the UIB, there are different reactions to water scarcity that are based on local conditions (Nüsser et al. 2019). Although not in every area of the region in terms of pace, intensity, or direction, the impacts on river flows may be significant.

Snow and ice melt as well as monsoonal precipitation regulate runoff in the lower Himalayas, all of which are anticipated to change due to climate change. The most sensitive parameter to climate change is monthly runoff as opposed to yearly runoff, especially in the spring and winter (Thayyen and Gergan 2009; Liu et al. 2012). Nepal and Shrestha (2015) reviewed studies published from 1996 to recent times to examine the impact of historical as well as projected changes in climate on the hydrological regime of three main river basins of Asia, namely, the Indus, the Ganges, and the Brahmaputra. They found that increase in both the temperature and precipitation can affect the hydrological regime of these river systems by modifying seasonal extremes, increasing evapotranspiration, altering glacier volume, and changing the snow and glacier melt. According to Kraaijenbrink et al. (2017), a 1.5 °C increase in global temperature will cause a warming of 2.1 ± 0.1 °C in the High Mountain Asia, and by the end of the century, 64 ± 7% of the present-day ice mass stored in the High Mountain Asia glaciers will still be present. The study's forecasts (RCP 4.5, 6.0, and 8.5) also show that a significant portion of the glacier ice is expected to vanish, with estimated mass losses of 49 ± 7%, 51 ± 6%, and 64 ± 5%, respectively, by the end of the century. The High Mountain Asia's regional meltwater output, however, is too high—at 1.6 times the equilibrium rate—and is expected to increase over the next several decades before eventually declining (Pritchard 2019). According to Zhao et al. (2022), the regionally averaged mass balance change rate in the High Mountain Asia was −21 ± 7 Gt/a from 2000 to 2021. By the end of the twenty-first century, total glacier mass will decrease by 46 ± 5% for Shared Socioeconomic Pathway (SSP) 119, 55 ± 3% for SSP245, and 70 ± 4% for SSP585 compared to 2000. The study also noted that, under all climate scenarios, 11 drainage basins' peak water may arrive before the middle of the twenty-first century, while the others would most likely reach peak water during 2030–2080. In tiny glaciers of the Trans-Himalayan Kang Yatze Massif, Ladakh, northwest India, Schmidt and Nüsser (2012) discovered a decrease (14% (0.3%/year)) from 96.4 to 82.6 km^2, and the average ice front retreat amounts to 125 m (3 m/year) between 1969 and 2010. There isn't much evidence in favor of significant decreases in the amount of water resources in the UIB, according to Archer et al. (2010). According to Mukhopadhyay and Khan (2014), climatic pattern changes will have a substantial impact on the UIB's water supply in the future. If glacial retreat and a decline in the area's perennial snow and ice cover occur owing to a changing climate, there will surely be long-term reductions in river flows in the UIB, which may jeopardize the sustainability of water supplies.

According to Bilal et al. (2019), a considerable increase in SCA above 5000 m ASL was seen in the UIB between 2000 and 2017. On the other hand, retreating glaciers also shield sizable populations from the effects of drought stress in the Asia's high mountains, where around 800 million people rely in part on glacial meltwater (Pritchard 2019). For the years 1993–2019, Jabbar et al. (2020) discovered a very minor rise in SCA in the UIB. From 2008 to 2018, the SCA decreased across the whole Indus basin and its sub-catchments, according to Ali et al. (2020). In their study of glacier fluctuations on the UIB's Nanga Parbat from 1856 to 2020, Nüsser and Schmidt (2021) discovered a striking resemblance between this glacier and the Karakoram's stable glacier mass. These modifications may have a negative impact on downstream communities and infrastructure, including the burgeoning hydropower sector and some of the largest irrigated agriculture systems in the world, by making water flow more severe and unpredictable. According to Wu et al. (2021), warming in the early twenty-first century resulted in glaciers in the Karakoram region of the UIB having a balanced or slightly negative mass budget of

−0.08 ± 0.07 m w.e. a^{-1} (inter precipitation and summer temperature). Debris-covered Hoksar Glacier (HG), according to Romshoo et al. (2022), is losing mass more quickly than a number of other Himalayan glaciers, as per the UIB continually negative mass balance data. Due to the presence of a heavy debris cover in the ablation zone, which routinely thins down toward the accumulation zone, the glacier demonstrated greater mass loss with elevation, as opposed to the typical decreasing mass balance with height. Increased black carbon concentration in the area, declining snowfall, and rising temperatures—all signs of climatic change—have all led to the increased mass loss of the HG. Depending on the scenarios and GCMs, Nazeer et al. (2022) demonstrate that the future UIB flow increase will range from 23% to 126% and the future glacier melt increase would range from 30% to 265%. The simulations indicate that all elevations will contribute more glacier melt than before, but the highest elevations will contribute by a significant amount.

The importance of glacier melt and snowmelt along with rainfall is indicated by Mukhopadhyay and Khan's (2014) findings that the contributions of glacier melt and snowmelt to annual river flows range from 18% to 35% and 38% to 50%, respectively, in the major tributaries and the UIB main stem depending on the region. If glacial retreats and a reduction in the amount of permanent snow and ice cover occur as a result of a changing climate, there will likely be long-term decreases in river flows in the UIB, which may jeopardize the sustainability of water resources in this basin. Azam et al. (2021) compiled the studies based on estimating the glacier/snow melt, rainfall, and total runoff in the HK region. According to the study, the contributions of glaciers and snow melt in various basins of the HK region range from 2% to 50% and from 22% to 65%, respectively. Since groundwater supplies have not increased to keep up with population growth, they have been overused in recent decades, which has caused groundwater levels in the Indus basin to drop (Akhter et al. 2021). According to Baig et al. (2021) glaciers account for two-thirds of the annual 28 river flows at the UIB, while precipitation and snowfall each provide 19 and 11%. The yearly river flows will fall by 16% overall in the UIB, with notable seasonal fluctuations. According to Latif et al. (2021), from 1968 to 2013, the UIB experienced a considerable increase in summer flow and a decline in August and September. According to Shah et al. (2020), the rise in annual average precipitation under RCP4.5 and RCP8.5 is 2.4% to 2.5% and 6.0% to 4.6% (mid- to late century), respectively. For the middle and late ages, RCP4.5 indicated a rise in the flow of 19.24% and 16.78%, respectively. According to RCP8.5, the flow will increase by 20.13% and 15.86%, respectively, in the middle and late of the century. According to Liaqat et al. (2022), over the period 1975–2017, the annual discharges at the Shyok, Gilgit, and Indus gauges in Kachura are marginally considerably increasing. They were also found to be more pronounced at the seasonal scale than at the annual scale in the UIB. By the end of the twenty-first century, Wijngaard et al. (2017) predict that future mean discharge and high-flow conditions will likely increase the UIB, Ganges, and Brahmaputra River basins. According to Ashraf and Rehman (2019), there was a moderately favorable association between mean temperature and rainfall in the UIB and river discharges from the sub-basins. According to Kiani et al. (2021), the UIB's October and April predicted inflow increases are the largest (37.99% and 65.11% at 1.5 °C and 2.0 °C, respectively), under RCP4.5 and RCP8.5, respectively. Increases in the snow and glacier melt contribution, which are more prominent at a warming level of 2.0 °C in the UIB, are causing these hydrological changes (Ougahi et al. 2022). There was a projected increase in summer precipitation (RCP8.5: +36.7%) compared to the baseline (1974–2004) and a decrease in winter precipitation (RCP8.5: −16.9%), with an increase in the average annual water yield from the glacial-nival regime and river flow peaks occurring 1 month earlier in the UIB. Khan et al. (2020) estimate that the UIB will experience yearly flow increases throughout the twenty-first century, but, curiously, these gains are greater in the middle years (2041–2070) than at the end of the century (2071–2100).

In the catchment region comprising the Chhota Shigri Glacier (WIH), Engelhardt et al. (2017) discovered that over the years 1951–1999, the simulated average annual runoff did not change significantly. A move toward earlier snowmelt beginning in 2040, however, would boost runoff in the summer months (May and June), while a decline in glacier melt would result in less runoff in the monsoon months (August and September). Tewari et al. (2017) analyzed the Satluj river basin's streamflow estimates for the mid-century under two scenarios (RCP 4.5 and RCP 8.5) and discovered how warming will affect streamflow. According to their projections, Satluj's overall annual discharge will decrease with time, particularly during the peak discharge season (JJAS). Sen and Kansal (2019) found that the communities think that climate change, which alters precipitation and temperature conditions, is to blame for the worsening of the spatiotemporal disparities in the Himalayas. As a result of the shift in melting, they also emphasized variations in the seasonal water supply. By incorporating all stakeholders, from local communities to the government, these challenges may be tackled using a variety of solutions in various areas. According to Jasrotia et al. (2021), precipitation forecasts have a significant impact on the Jhelum catchment's runoff in the WIH. From 2020 to 2080, runoff gradually increases, and then it gradually declines.

Mishra and Lilhare (2016) concerned with the hydrological sensitivity of 18 major river basins of the Indian sub-continent, with special focus on Ganga and Godavari basins, to climate change during the twenty-first century found that most of the river basins are likely to get wetter and warmer in future, as a result of continuously rising air temperature and increasing precipitation, though there is a large amount of uncertainty in both the spatial and temporal precipitation patterns (Mishra and Lilhare 2016). In addition, research on the Ganga and Godavari basins, two significant river basins in the north and south, suggests that surface runoff and streamflow are more sensitive to air temperature and precipitation than ET and that changing precipitation patterns, rather than the air temperature, primarily control these processes. In the Ganges basin, it is predicted that a 30% drop in precipitation combined with a 4% increase in air temperature may result in a decline in surface runoff and stream flow of 55%, while comparable climatic scenarios may result in a decline of more than 60% in the Godavari basin (Mishra and Lilhare 2016).

Rani et al. (2019), in their study based on the upper Beas basin of the WIH, found that the future climate scenarios, i.e., increasing temperature, increasing rainfall, and rapid melting of snow, may cause PET to increase by nearly 2% in the basin by 2050s, with significant seasonal variations. Further, the study suggests that PET, being much more sensitive to changes in temperature, increases with rising temperature and in turn leads to a rise in crop water requirement, but it is compensated by enough water being available due to early and rapid melting of snow. Another study conducted in the same basin suggested that the basin is likely to get warmer and wetter in comparison to the historical period because of increasing temperature as well as precipitation till the end of the twenty-first century, with the change in precipitation being more uncertain than that in temperature (Li et al. 2019a, b). The increase in temperature and precipitation leads to accelerated glacier ablation and a slight increase in streamflow or runoff in the future, although there is significant seasonal variation and a large amount of uncertainty. In comparison to 2005, it is predicted that the glacier extent of this basin would decline by roughly 63–87% in the middle of the century and by 89–100% at the end of the century. This rise in precipitation and decline in snow cover or glacier extension point to a complex future picture of total discharge with an augmentation of the winter and pre-monsoon period, but it is challenging to make any predictions with accuracy regarding the monsoon season (Li et al. 2019a, b). The intensified water cycle and extreme rain events are most likely to affect the hydrological behavior of a region, causing extreme flooding in some and intense droughts in other areas. Moreover, the precipitation patterns are expected to change spatially as well as temporally. It is also noteworthy that mountainous areas prove to be more sensitive to climate change because they have immense storage of water in the form of snow and glaciers at high altitudes and are home to the headwaters of most of the extensive river networks worldwide (Xenarios et al. 2019). Any slightest variation in climate over the mountains may have dreadful consequences, not only for the people directly dependent on them for their livelihood but also for the people living in downstream areas, depending indirectly on these mountainous areas for various resources—freshwater is the most invaluable one.

Studies found that changing LULC, i.e., rapid urbanization, infrastructural developments, deforestation, etc., may cause an increase in the effective impervious area and a decrease in the forest cover. As a result, there would be an increase in total discharge and surface runoff leading to reduced water availability for ground infiltration and soil moisture replenishment, which in turn causes the groundwater recharge and baseflow to decrease. Moreover, deforestation may cause the water retention capacity of watersheds to decrease and the increased impervious area may lead to increased flood peaks during storm events in the future (Talib and Randhir 2017). A study conducted in the Wei River basin of China, which is highly modified due to anthropogenic activities, suggested that the LULC changes may lead to a reduction in streamflow of the watershed, particularly during the dry season (Li et al. 2019a, b). The main reason behind the declining annual and dry seasonal streamflow is the large reduction in surface water due to the expansion of cropland in the region in the last 30 years. However, the loss of forests or woodlands decreases the soil water, which in turn leads to a meager increase in streamflow in the dry season. Besides this, the decrease in streamflow during the wet season is caused due to decreased groundwater (Li et al. 2019a, b). Wang-Erlandsson et al. (2018) in their study found that human-induced LULC changes have resulted in a reduction of evaporation and precipitation and caused river flows to increase in large parts of the world.

Even a slight change in LULC, due to socioeconomic and infrastructural developments taking place in a region to meet the needs of its huge population, may affect hydrological behavior of a river basin in a greater way (Pokhrel et al. 2018), especially the developing countries that are characterized by the agrarian economy, like India, because the unprecedented growth in their population imposes huge pressure on their LULC pattern (Garg et al. 2019). For instance, the already devastating flood of September 2014 that occurred in the Jhelum basin of Kashmir in India further became disastrous because of anthropogenic factors like large-scale rapid urbanization, infrastructural developments (e.g., railways and highways), loss of wetlands, deforestation, decrease in drainage capacity due to the siltation of water courses, etc., suggesting that the changes occurring in LULC in a region can exacerbate the adverse effects of changing climate on hydrological processes of the region (Romshoo et al. 2018). Some studies (Dame et al. 2019; Müller et al. 2020) expressed concerns about how such unchecked urban growth and rapid urbanization in Leh (Ladakh, Upper Indus basin) would affect urban and environmental governance, particularly in relation to water resources and natural disasters and provide useful guidance for sustainable town planning. Another study conducted by Garg et al. (2019) in India, particularly Pennar river basin, located in the south, to analyze the potential impacts of anthropogenically induced changes in LULC on the hydrology of a river basin, shows an increase of nearly 45% in average annual runoff of the basin between 1985 and 2005 due to the changing LULC scenarios such as an increase in urban sprawl by 0.14% at the expanse natural forest area and cropland, which influences ET in the region and hence the other hydrological components also get affected, impacting the water availability in the region. The baseflow, also, shows a slight increase due to a rise in the number of water bodies in the basin during this period. Similarly, Wagner et al. (2016) examined the probable effects of LULC modifications on water resources of a rapidly developing watershed upstream of Pune, located in the Western Ghats, India, in the near future (i.e., between 2009 and 2028) and found that altered land use (such as increasing urban spread and declining area under agricultural land and natural vegetation) leads to seasonal variations in water balance components. The most prominent ones to be expected in the near future (2009–2028) are an increase in water yield in the beginning of monsoon by nearly 11 mm/month caused due to increase in impervious area (growth of about 23.1% in urban area) and decrease in evapotranspiration in the dry season by about 15.1 mm/month as a consequence of a change in an irrigated agricultural area (a decline of nearly 14%) (Wagner et al. 2016).

Anand et al. (2018), in their assessment of hydrological behavior of Ganga River basin with changing LULC, found that urbanization and subsequent deforestation are the main contributing factors toward increasing surface runoff and water yield in the basin, while the increased irrigation requirements lead to water consumption and an increase in the process of evapotranspiration. Thus, it is evident altered land use practices and changes in the distribution of vegetation in a region largely impact water balance components and flow regimes of the river basin because different land cover layers have a different impact on basin hydrology. Moreover, the interrelationship between soil and vegetation can modify the hydrological cycle at a local level, impacting the yearly distribution of streamflow in the basin (Anand et al. 2018). A scientific report, assessing the sensitivity of Indian summer Monsoon Rainfall (ISMR) toward the large-scale deforestation that has taken place in our country, suggests that the ISMR gets weakened as a result of deforestation because it leads to a decrease in evapotranspiration, which in turn causes a reduction in the recycled component of precipitation (Paul et al. 2016).

Studies suggest that the combined impacts of climate change and LULC changes may prove to be more profound than their impacts and are expected to increase total runoff, surface runoff, and average storm peak values (Neupane et al. 2015; Morán-Tejeda et al. 2015; Talib and Randhir 2017). Some studies suggest that river flows may get affected by both the LULC changes within the basin and those occurring outside the basin through atmospheric teleconnections (Wang-Erlandsson et al. 2018). This implies that local land use changes observed in one part of the world can also be linked to changes in remote precipitation occurring far away, which in turn affects distant river flows, through land-atmosphere interactions such as terrestrial moisture recycling, import and export of atmospheric moisture, etc. Thus, a proper understanding of the land and atmosphere interactions is necessary for analyzing the changes in river flows and water resource management and for adapting to this rapidly changing world, both in terms of climate and LULC.

Zhou et al. (2015) tried to theoretically analyze the existing global pattern and compared the roles of climate variations and land cover changes in affecting the hydrological processes, which indicates the relation of water yield with wetness index and watershed characteristics. This study based on three parameters, namely, precipitation, energy or potential ET, and watershed characteristics, shows that climate change plays a more important role in humid regions and watershed areas having high-water retention capacity, while land cover changes may cause greater and more sensitive hydrological responses in non-humid areas and watersheds characterized by low-water retention capacity. Hence, this study challenges the commonly held perception that hydrological responses are mainly governed by climatic variations, while land cover changes have a secondary role to play (Zhou et al. 2015). Neupane et al. (2015) studied the probable hydrological changes that may be observed as a result of changing climate and land use practices in the monsoon-dominated Himalayan basins, particularly the Tamor and Seti River basins located at the eastern and western flanks of Nepal, respectively. Under the influence of changing climatic scenarios, higher stream discharge is expected in both basins due to the increase in precipitation, while a reduction in the magnitude of stream discharge is likely to occur because of changing LULC scenarios. However, the study of their integrated impact shows a moderate increase in discharge as the reduction in discharge caused by land use changes will be compensated by the increment under changing climatic scenarios. Pervez and Henebry (2015), in their attempt to assess the availability of freshwater in the Brahmaputra Basin in the wake of projected changes in climate and LULC scenarios by the end of twenty-first century, found that these changes are likely to have a positive impact on the water resources of the Brahmaputra River basin, although worsened drought and flooding situations may occur because of probable decrease in total water yield, streamflow, and soil water content in May–July and projected rise in the seasonal increase in streamflow and water yield in August–October, respectively (Pervez and Henebry 2015). Although a sufficient amount of research has been done on the potential impacts of climate change and LULC changes individually, not much work has been done on their combined impact. Thus, there is a need to study the hydrology of a river basin using integrated approaches and models that are sensitive to both climate change and LULC (Zhou et al. 2015) as they can provide useful information about watersheds and will help us in making more effective plans and decisions for the management of the water and land resources (Dwarakish and Ganasri 2015). Haleem et al. (2022) demonstrate that climate change (61.61%) has a greater impact on river runoff in the UIB than land use change (38.39%) from 2000 to 2013. Both climate and land use changes are expected to result in deeper runoff in the future. Land use change has a smaller impact (0.37–1.1%) than climate change (12.76–25.92%).

In the upper Beas basin, the climate change scenarios (such as an increase in air temperature and rainfall, prolonged and accelerated snow-melting, etc.) lead to a substantial increase in discharge, while the land cover changes (like increase in an urban area and agricultural land, decrease in forest cover and snow cover, etc.) may cause a remarkable decrease in discharge (Rani and Sreekesh 2021). Moreover, the increase in discharge occurring as a result of climatic changes may not continue under changing land use, and also there may be a probable decline in discharge if climatic conditions continue to change at a similar rate. Therefore, these changes may initially be beneficial for economic activities but may lead to the serious problem of water scarcity in the longer run, considering the rapid rate of population growth and urbanization in the region. It also highlights the fact that integrated impacts of climate and LULC changes hold greater significance than their individual effects.

1.5 GIS Models to Assess the Impact of Climate and LULC Changes on Hydrology

Without a reliable hydrological model, it is impossible to assess the impact of climate and land cover changes on basin hydrology. Over the last four decades, computer-based hydrological models have been created and used at various scales for a variety of applications. There are two main causes behind this. For starters, new models and approaches are always being developed by the research community. Second, as the need for improved tools grows, so does the pressure on water supplies. There are a variety of hydrological models available, ranging from pure empirical techniques to dynamic three-dimensional models that run on a geographic information system (GIS) platform (Table 1.1). There are various advantages of using hydrological models for impact assessment studies:

These models are set up for use at various geographic scales and have been validated for a range of climatic and physiographic conditions.
These could help meet the characteristics of accessible data. Simulated hydrological responses to many scenarios of changing the temperature and land cover are possible.
Compared to general circulation devices, these variants are a great deal more user-friendly.
These may be used to determine how sensitive specific watersheds are to real-world climate and land cover changes as well as those predicted by large-scale models.
It will be beneficial to use techniques that can incorporate information from large-scale models as well as specific area hydrologic characteristics.

In the present study, the Soil and Water Assessment Tool (SWAT) model is found suitable for evaluating the impact of climatic and land cover changes on runoff. Arnold developed the SWAT, a basin or watershed size model for the Agricultural Research Service (ARS) of the United States Department of Agriculture (USDA) (Neitsch et al. 2002). SWAT has been used in studies all around the world to assess the effects of climate change on river runoff at both the global and regional levels (Nilawar and Waikar 2018; Osei et al. 2019; Zhang et al. 2020). The SWAT model offers various benefits over other models, including the following:

The model integrates empirical and physically based equations makes use of easily available inputs and allows users to investigate long-term effects.
It is a distributed model with a daily time step that operates on a continuous basis.
To characterize the connection between variables of entrance and exit, certain precise information on weather conditions, soil, geography, vegetation, and management is required.
Its validity has been evaluated for a variety of basin sizes throughout the world.

1.6   Statement of the Problem

Rivers in Himachal Pradesh are the main source of water for domestic, commercial, and industrial usage. Evidence of changes in precipitation (rain and snowfall) in the state indicates that the intensity and frequency of snowfall and rainfall have decreased over time, leading to a decrease in the amount of water available in the state's rivers (Government of Himachal Pradesh 2002a, b), though the precise amount of river flow reduction is unknown. Since the micro-scale analysis takes into

Table 1.1 Hydrological models

Name		Developers	Description
SRM	Snowmelt-Runoff Model	Martinec (1975)	Simulate the hydrological response on snow and glacier melting runoff
HBV	Hydrologiska Byråns Vattenbalansavdelning model	Harlin (1991)	Analyze river discharge and water pollution
PRMS	Precipitation-Runoff Modelling System	Yan and Haan (1991)	Evaluate the response of streamflow under various combinations of climate and land use
CASC2D	CASCade of planes, two-dimensional	Julien et al. (1995)	Simulates Hortonian overland flow (HOF) in a watershed
SWAT	Soil and Water Assessment Tool	Arnold et al. (1998)	To quantify the impact of land management practices in large, complex watersheds
WASMOD	Water And Snow balance Modeling system	Schimming et al. (1995)	Snow accumulation and melt and actual evaporation separate runoff into a fast and a slow component with a monthly time step
DHSVM	Distributed Hydrology-Soil-Vegetation Model	Nijssen et al. (1997)	Distributed hydrologic model that explicitly represents the effects of topography and vegetation on water fluxes through the landscape
SWIM	Soil and Water Integrated Model	Krysanova et al. (1998)	Simulating runoff generation, nutrient and carbon cycling, plant growth and crop yield, river discharge, and erosion as interrelated processes at a daily time step
MIKE SHE	Generalized River Modeling Package-Systeme Hydraulique Europeen	Bøggild et al. (1999)	Advanced, physically based, distributed, continuous hydrologic and hydraulic simulation model
HMS	Hydrologic Modelling System	Yarnal et al. (2000)	Design of drainage systems, quantifying the effect of land use change on flooding
MONERIS	Modelling Nutrient Emissions in River Systems	Behrendt and Opitz (2000)	Quantification of nutrient emissions from point and diffuse sources in river catchments
WASIM	Water balance Simulation Model	Rode and Lindenschmidt (2001)	Spatial and temporal variability of hydrological processes in complex river basins
J2000		Krause (2002)	Simulation of hydrologic processes
DWSM	Dynamic Watershed Simulation Model	Borah et al. (2004)	Simulations of surface and subsurface storm water runoff Propagation of flood waves, soil erosion, and entrainment and transport of sediment and chemicals (nonpoint source pollutants)
MIKE BASIN		Ireson et al. (2006)	Steam and groundwater hydrology and water quality

consideration the local factors that influence changes in hydrological variables, a larger-scale research that does not account for spatial variability may distort the genuine trend analysis. The Kullu District in Himachal Pradesh, where surface water serves as the main source of water, may potentially experience the effects of global warming (Beas River). The Himachal Pradesh State Climate Change Action Plan (HPSCCAP) (GoHP 2012) states that as temperatures have risen, rainfall has replaced snowfall on the upper slopes of Kullu, and rainfall intensity has increased throughout the area. The Beas River discharge at the Bhuntar, Thalout, and Pandoh stations has significantly decreased, according to several studies (Bhutyani et al. 2008; Kumar et al. 2009; Singh et al. 2014). The snow and glaciers in the area provide water for the rivers. In the Kullu District of Himachal Pradesh, where surface water is the main source of water, the effects of warming may also be noted (Beas River). The high slopes of Kullu have received rainfall rather than snowfall due to rising temperatures (GoHP 2012), and rainfall intensity has increased in the area. A significant decrease in Beas River discharge has been seen at the Bhuntar, Thalout, and Pandoh stations (Bhutyani et al. 2008; Kumar et al. 2009; Singh et al. 2014). The snow and glaciers in the area supply the rivers in the area with water.

A total of 47 snowfields has an aerial extent of 72.4 km^2, along with 6 glaciers that have a surface area of around 16 km^2. Additionally, the size of these glaciers has declined (Kulkarni et al. 2005; Kumar et al. 2009; Dutta et al. 2012). Between 1963 and 2000, the Beas Kund glacier, which is the source of the Beas River, showed a depletion of 18.8 m/year, and monsoon flow in the basin showed a notable reduction (GoHP 2012). Based on population projections, the Human Development Report of HP (GoHP 2002a, b) states that in the district's urban areas, water consumption will increase from 2.99 lpcd (liters per capita per day) in 1999 to 4.82 lpcd in 2021. In rural areas, water use would increase from 23.61 lpcd in 1999 to 38.89 lpcd in 2021. Due to the district's high concentration of rural residents, it shows that the water demand has increased more in rural areas than in urban ones. Water stress would arise from the district's population being more dependent on surface water sources due to rising water demand and declining water supplies.

These situations suggest that the hydrological cycle has been significantly impacted by changes in the climate and land cover. To predict potential future changes in water resources and support local water management, it is also crucial to analyze the impact of climate and land cover changes on streamflow of the basin. A survey of the literature revealed the lack of in-depth and updated studies that investigate the connections between temperature, precipitation (snow and rain), changes in land cover, and streamflow. Furthermore, it has been emphasized that although climate change is a worldwide phenomenon, its severity and effects vary on different time and space scales. Additionally, the flow is being altered by changes in land cover. The goal of the current study is to provide a comprehensive examination of the impact of potential changes in the climate and land cover on streamflow. It also looked into the possibility that changes in the land cover and climate (including temperature and precipitation) of the basin were affecting streamflow. What sort of modifications, if any, occurred? The study will be useful for effective water resource planning and management to keep people's livelihoods upstream and downstream.

The following are the objectives of the present work:

To look at the monthly, seasonal, and yearly variations in mean air temperature (T_{mean}), ET, cloud cover (CC) (low, middle, high, and total), cloud base height (CBH), precipitable water vapor (PWV), and rainfall

To understand the changes in LULC in the basin

To investigate the impact of future climate change on streamflow in the basin by 2071

To evaluate the impact of projected land cover changes on streamflow in the basin by 2071

1.7 Study Area

This work is based on the Beas River, also known as Vipasha in Sanskrit literature and Arjikiya in Vedic literature. The Beas River is one of the principal tributaries of the Indus River and the principal river of India. This basin is located in India's lower Himalayan region between the latitudes of 31° 7′ and 32° 33′ North and the longitudes of 74° 54′to 77° 51′East (Fig. 1.1). It occupies roughly 5.63% of the Indus River basin's surface (India-WRIS 2012). It begins at the Beas Kund, which is located at an elevation of 4085 meters above mean sea level (AMSL), close to the Rohtang Pass in the Kullu district of the Himachal Pradesh state (Kathuria and Thakur 2004). It is the second most important river in the state of Himachal Pradesh

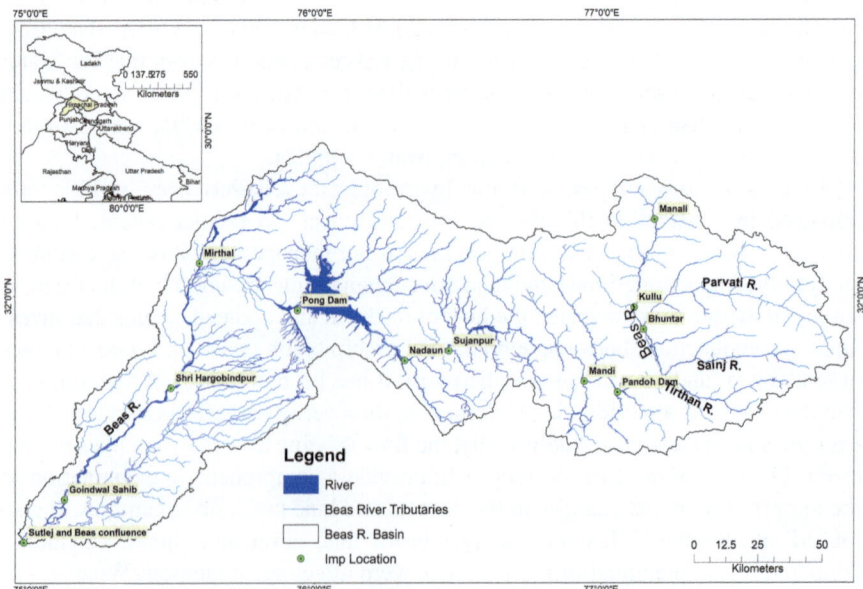

Fig. 1.1 Location and extent of the Beas River basin with the drainage system

and the second easternmost of the rivers of Punjab in India. It is a small stream at the source but receives more water from the spring during its flow path. Subsequently, it flows south through the Kullu valley (join by Parvati, Sainj and Larji rivers), nearly north-south direction up to Larji (the first Larji dam near Aut with a reservoir with a catchment area of 4921 km²).

After the Parvati River and the river's confluence at Bhuntar (a town in the Himachal Pradesh state), the river's flow is approximately twice as high (Kathuria and Thakur 2004). It runs in the same way up till Pandoh Dam (built on the Beas River in the Kullu region of Himachal Pradesh), when it makes a right angle and heads southwest (the second dam on the river with a reservoir with a catchment area of 4921 km²). The Beas River runs for 116 km till it reaches the Pandoh dam. Only 780 km² of the 5300 km² catchment area of the Beas River basin up to the Pandoh Dam is permanently covered in snow (Bhakra Beas Management Board (BBMB) 1988). The water from Pandoh Dam has been redirected to join the Satluj River at Slapper through a large tunnel that runs through the open Sundernagar Canal, which is a town and a municipal municipality in the Mandi district of the Indian state of Himachal Pradesh (also known as Beas Satluj link).

Then it turns west to flow past Mandi into the Kangra valley and join by Uhel river near Mandi town. It covers a length of about 256 km with an area of 13,663 km² in Himachal Pradesh before it enters the Punjab plains at Mirthal. It covers the districts of Kullu, Mandi, Chamba, Una, Hamirpur, Kinnaur, Lahul and Spiti, Shimla, and Kangra in Himachal Pradesh. It has another dam at this location which is known as Pong dam (also known as the International Ramsar site and Maharana Pratap Sagar with a catchment area of 4921 km²). It covers the districts of Amritsar, Firozpur, Gurdaspur, Hoshiarpur, Jalandhar, and Kapurthala in Punjab state. It meets the Satluj River near Harike (south of Amritsar in Punjab) at around 460 km from the source. The river's catchment area is 20,303 km², and its whole length is 460 km (from the source to the destination).

1.8 Data and Sources

Secondary and primary data were used in the present study. Fieldwork was done to evaluate the accuracy of digital elevation models (DEMs) and LULCs, as well as to gather data on LULC-related parameters needed for the SWAT model. The study included remote sensing, soil type, and climate data (Table 1.2).

1.8.1 Meteorological Data

ERA5 reanalysis PWV (mm, 850 hPa), air temperature (°C at 2 m), rainfall (mm), CBH (m), CC (%), and ET (mm) at $0.25° \times 0.25°$ horizontal resolution were obtained for the period 1980–2020 (http://climate.copernicus.eu/climate-reanalysis)

Table 1.2 Data and their sources of the present work

Data	Sources
Mean air temperature (T_{mean}), evapotranspiration (ET), cloud cover (CC) (low, middle, high, and total), cloud base height (CBH), precipitable water vapor (PWV), and rainfall (1980–2020)	Copernicus Climate Change Service (C3S) http://climate.copernicus.eu/climate-reanalysis
Discharge of Thalout station (1971–2003)	Ghorpa Hydal project report (www.powermin.nic.in)
Topographic map at 1:50,000	Survey of India (SoI)
Digital elevation model (DEM) 30 m	Cartosat-1 Dem http://bhuvannuis.nrsc.gov.in/bhuvan/web/
Snow cover area (MOD10A2) 2001–2020 (500 m)	The Moderate Resolution Imaging Spectroradiometer (MODIS), Data Pool at National Snow and Ice Data Center (NSIDC), Distributed Active Archive Center (DAAC)
Vegetation monthly data (NDVI layer) (MOD13A1) 2020, 250 m	USGS Earth Explorer
Satellite images	USGS Earth Explorer
Soil map Type of soils and their chemical properties	Soil Resource Mapping District Kullu and Mandi, Himachal Pradesh, 2013, Soil and Land Use of Survey India (SLUSI)

(Copernicus Climate Change Service (C3S) 2017; Hersbach et al. 2018). For the SWAT model, temperature and rainfall data between 1969 and 1980 are filled by IMD gridded data. ERA5 is a fifth-generation European Centre for Medium-Range Weather Forecasts (ECMWF) atmospheric reanalysis data of the global climate covering the period from January 1950 to the present. Daily and monthly aggregates of ERA5 data are available which are produced by the C3S at ECMWF. A variety of ground-based and satellite measurements are combined to provide an atmospheric reanalysis dataset that is based on an AGCM (Zhang et al. 2013; Chen and Liu 2016). These data have been used extensively in climate change studies because of their several advantages over other datasets such as homogeneous records, global coverage, spatial integrity, and multivariable outputs (Allan et al. 2002; Lu et al. 2015).

1.8.2   Terrain Data

Cartosat-1 Dem 30 m was downloaded from Bhuvan[2] for demarcation and morphometry analysis of the Beas basin. The data was also needed to produce a decision tree for LULC mapping of the study area. The Survey of India (SoI) provided the toposheet Nos. 53E/2 (1970), 53E/6 (1974), 143X12 (2004), 143X8 (2004–2005), 143X7 (2004), 143X3 (2004), and 143X4 (2004) at 1:50,000 for the preparation of

---

[2] http://bhuvannuis.nrsc.gov.in/bhuvan/web/

the base map of the basin and LULC accuracy evaluation. To create the basis map for LULC mapping of the Beas basin, toposheets from 1964 to 1965 (53 E1, 53 E2, 53 E9, and 53 E13) and 2005 to 2006 (52 H3, 52 H4, and 52 H7) at 1:50000 were downloaded from the SoI.

1.8.3 Land Use/Land Cover Data

Earthdata Search has provided the Moderate Resolution Imaging Spectroradiometer (MODIS) MOD12Q1 Land Cover Type version 6 data for the years 2001 and 2020 (Table 1.3) (Friedl and Sulla-Menashe 2019). The European Commission's GHSL (Global Human Settlement Layer) database (https://ghsl.jrc.ec.europa.eu/datasets.php) has provided multi-temporal classification of the built-up presence of GHS BUILT LDSMT at 30 m (derived from Landsat, multi-temporal) and GHS SMOD POP2015 population density (derived from GPW4.10 at 250 m) of the basin (Freire et al. 2016; Corbane et al. 2018, 2019; Florczyk et al. 2019; Schiavina et al. 2019). To create the LULC map of the study area, path-147 and row-38 of the post-monsoon Landsat images of October/November were obtained (Table 1.3). MODIS data were used for a time series analysis of SCA.

1.8.4 Soil

The study area's soil map and its characteristics (physical and chemical) were taken from the Soil Resource Mapping District Kullu and Mandi, Himachal Pradesh, Soil and Land Use of Survey India (Table 1.2) (SLUSI 2013).

Table 1.3 Details of satellite images used in LULC analysis

Satellite data	Sensor	Path/ row	Date of acquisition	Number of bands
MODIS MOD12Q1			2001–2020	1
Global Human Settlement Layer Built-up area grid (GHS-BUILT) derived from Landsat (Chapter 1 material and method section)			1975–2014	1
GHS Settlement Model grid (GHS-SMOD)			1975–2015	1
LANDSAT_3	MSS	158/038	22/10/1980	4
LANDSAT_7	ETM+ SLC on	147/038	15/10/2000	8
LANDSAT_8	OLI_ TIRS	147/038	15/10/2020	11

Note: *MSS* multispectral scanner, *TM* thematic mapper, *OLI_TIRS* Operational Land Imager and Thermal Infrared Sensor, *ETM+SLC On* Enhanced Thematic Mapper Plus

1.9   Methods

1.9.1   To Analyze the Climate Variability of the Study Area

Obtained climate data has been clipped to the study area in GIS. All the collected parameters were aggregated on a monthly scale, and its trend analysis has been done using the ordinary least square regression test, and significance is tested at 0.05 and 0.01 levels (Fig. 1.2).

1.9.2   To Understand LULC Changes in the Basin

1.9.2.1   Morphometric Analysis

Morphometric analysis of the upper and middle Beas basin has been done in the present study because the DEM of the lower parts of the river is of poor quality (Fig. 1.3 and Table 1.4). Cartosat 1 DEM images have been mosaicked in the GIS environment and reprojected into Universal Transverse Mercator (UTM) 43° N. Morphometric analysis of a river basin gives a realistic picture of the current geo-hydrological condition, nature of the rocks form, permeability, and porosity of the soil (Sarkar et al. 2020). Six parameters from the linear aspects, seven parameters from areal aspects, and six relief aspects were analyzed (Fig. 1.3 and Table 1.4). These different parameters can be interpreted with the help of remote sensing and GIS (Singh et al. 2013).

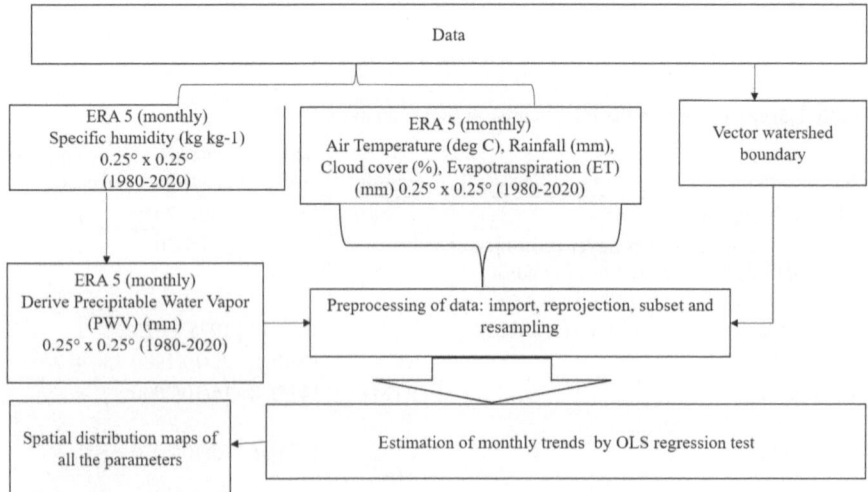

Fig. 1.2 Method of analyzing the climate variability in the present study

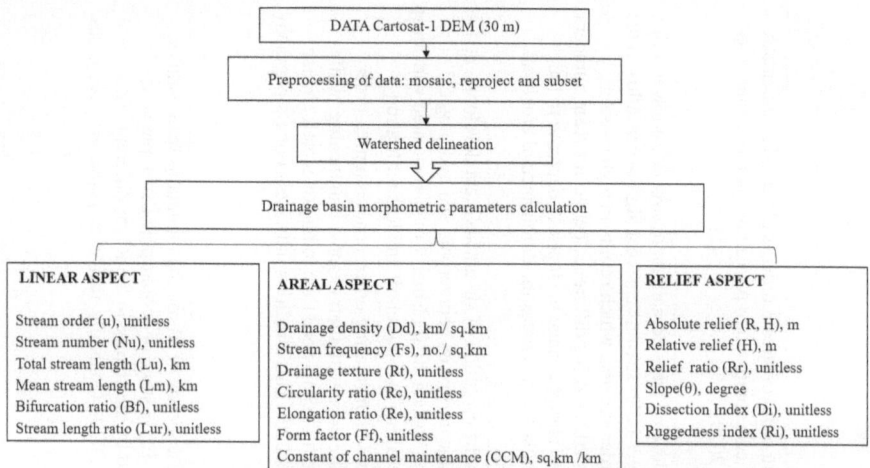

Fig. 1.3 Method of morphometric analysis of the Beas River basin in the present study

1.9.2.2 LULC Analysis

LULC analysis of the entire basin was done using MODIS-processed products. The LULC of the upper Beas basin has been done using Landsat images for more accuracy in the SWAT model. First, the upper Beas River basin region was extracted from the satellite photos using the QSWAT model, and the basin border was marked using the Cartosat DEM 30 m in QGIS. The study calculated the images' top-of-atmosphere (TOA) reflectance. The metadata of the images was used to determine gains, offsets, solar irradiance, sun elevation, and acquisition time (Fig. 1.4). To remove the dark pixel values from the images, dark object subtraction (DOS) atmospheric correction was used. To make the MSS Landsat image compatible with other Landsat images for creating the LULC transition/conversion matrix, the image's resampling from 60 m to 30 m was performed.

The SoI toposheet was used to create a base map of the study area showing the LULC classes in the basin. Seven LULC classes were defined in the study, following the National Remote Sensing Centre (NRSC) and Indian Space Research Organization (ISRO) (2011). The LULC categorization scheme using a decision tree classification approach (Figs 1.5 and 1.6) is as follows:

Built-up area
Cultivated land
Forest
Grassland
Barren/unculturable/wasteland (BUW)
Water bodies
Snow

Table 1.4 Morphometric parameters used in the present study

Parameters	Formulas	Unit	Description	References	Remarks
Order of stream (u)	Hierarchical ordering	Unitless	Hierarchical ranking using Strahler's method of ordering	Horton (1945), Strahler (1952, 1964), and Smith (1950)	Increasing order of stream order is observed in mountain environments compared to plateau or plain environments (Mahala 2020)
Stream number (Nu)	Number of stream of a particular order (Nu)	Unitless	Counting of particular order Stream one by one	Horton (1945) and Strahler (1952, 1964)	Highly elevated areas contain more streams of lower order. More lower-order streams ensure a high-water influx in the higher-order streams, which results in high erosion (Vishwas 2021). A big difference between stream numbers denotes an abrupt change in slopes. A decrease in the number of streams denotes the dominance of erosional landform
Stream length (Lu)	Length of a stream of a particular order	Km	Stream length of lower order will be more especially in the mountain region	Horton (1945)	Indicates the evolution of stream segment development. Direct indicators of its geomorphic and hydrological sequence (Mahala 2020). Small Lm is associated with steep slopes resulting in high runoff and less infiltration
Mean stream length (Lsm)	Lsm = Lu/Nu, where Lu = total length of streams (km) of a particular order "u," Nu = Total number of streams of a particular order "u"	Km	Total stream length of an order divided by the total length of the same order. Decreases with stream order	Horton (1945)	Increases with increasing order, denotes higher-order streams are done with their channel lengthening, while lower-order streams need time to lengthen their channel (Mahala 2020). This condition indicates young topography
Stream length ratio (RL)	RL = Lsm/(Lsm−1) where, Lsm = mean stream length of a particular order 'u', Lsm−1 = mean stream length of next lower order 'u−1'	Unitless	Ratio between Lsm of one order to the next lower order	Horton (1945)	Irregular tendency of RL on the mountains than plain or plateaus. Early stage of geomorphic development is suggested by the change in the RL of different orders (Mahala 2020). Generally, increases as the order increases

	Formula	Unit	Description	References	
Bifurcation ratio (Rb)	$Rb = Nu/Nu + 1$ Nu = number of streams of a particular order "u," $Nu + 1$ = Number of streams of the next higher order "u+1"	Unitless	Number of streams of one order divided by the streams of next higher order	Horton (1945), Strahler (1957), Maxwell (1955), and Schumm (1956)	The measure of the flood potentiality for any river basin. Streams on the high undulated terrains are chiefly the product of the lithological controls of the surface. High Bf on mountains denotes stream network is highly impacted by the lithology and geology of the basin
Drainage density (Dd)	$Dd = \Sigma L/A$ where ΣL = total length of streams (km) and A = basin area (km²)	km/km²	Ratio of total stream length to the total area of the basin	Horton (1945)	Measures surface runoff and dissection character quantitatively (Chorley 2019). Climate and vegetation are responsible for the density of streams in an area (Moglan et al. 1998). Dd is high in high-relief areas, related to more runoff, low permeability, and high gradient; the opposite of this is observed in low Dd (Nag and Chakraborty 2003)
Stream frequency (Fs)	$Fs = Nu/A$ where Nu = total number of streams in a drainage basin, A = basin area	No./km²	Ratio of the number of streams in a unit area to the total area of the basin	Horton (1945)	Infiltration capacity, relief, and permeability is judged. High Fs in mountains denote low infiltration and more runoff due to high slopes, and the opposite is in the case of plateau plain areas
Texture ratio (Rt)	$Rt = Nu/P$, where Nu = Total no. of streams and P = perimeter of the basin	Unitless	Ratio of total no. of streams to perimeter of the basin	Horton (1945)	Some influencing factors of Rt are rainfall, vegetation, infiltration capacity, type of soil, relief, and geomorphic development stage (Smith 1950). The coarse texture indicates low Dd with high infiltration, whereas the fine texture indicates high "Dd" with more runoff
Circulatory ratio (Rc)	$Rc = 4\pi A/p2$ where A = area of the basin (km²) and P = outer boundary of a drainage basin (km)	Unitless	Area of the basin divided by the area of the circle with the same circumference as the basin perimeter	Miller (1953)	The value near "0" indicates less circular or highly elongated shape, homogenous and permeable geology, and the value near "1" indicates highly circular or less elongated shape, heterogeneous, and less permeable geology. High peak flow with a shorter duration of time is implied by high Rc values with more structural disturbances or structural control. Generally, mountain–plain fronts have high Rc value or form circular basins than the plain-plateau fronts which have elongated basin shapes

(continued)

Table 1.4 (continued)

Parameters	Formulas	Unit	Description	References	Remarks
Elongation ratio (Re)	Re = $(2\sqrt{(A/\pi)})/L$ or Re = $P/\pi L$; P = outer boundary of a drainage basin (km) L = basin length (km)	Unitless	Diameter of a circle having the same area as the basin area divided by the length of the basin	Schumm (1956)	Its values also range between "0" and "1." A low numeric value means elongated shape, whereas a more numeric value means circular shape. Revalue of "1" is a low-relief region; the range between 0.6 and 0.8 is a high relief with a steep slope region (Strahler 1957)
Form factor (Ff)	Ff = A/L^2 where A = area of the basin (km^2) and L = basin length (km)	Unitless	Total area of the basin divided by the square of the axial length of the basin	Horton (1932)	Form factor is used in identifying basin shape and related hydrological characteristics, sediment transport, defining flood corridor, and flood formation. The 'Ff' value varies from "0" to "1." A value near "0" indicates an elongated shape, low sediment transport rate, and low probability of flooding in lower course and low-relief areas because it takes a longer time to reach the main river (Vishwas 2021)
Constant of channel Maintenance (CCM)	CCM = 1/Dd where Dd = drainage density	km^2/km	Minimum area required for the maintenance and the development of the channel. It is the reciprocal of the Dd (Schumm 1956)	Schumm (1956)	A low CCM value indicates high flood potentiality and young geomorphology (Mahala 2020), and low CCM is the character of the high-relief areas due to low infiltration, high runoff, and high-drainage density, whereas a high CCM value indicates low flood potentiality and mature to old geomorphology, and high CCM is observed in low-relief areas having high infiltration, low runoff, and low-drainage density
Absolute relief (R)	Maximum height	Meter	Maximum vertical range in a basin		It is the maximum height of an area
Relative relief (H)	H = R−r, where R = highest relief and r = lowest relief	Meter	It is the difference between the maximum and the minimum elevation in the region	Schumm (1956)	The difference between the maximum and minimum relief. Useful in calculating most of the relief aspects of the basin

Relief ratio (Rr)	Rr = (H/L) where H = relative relief (m) and L = length of the basin (m)	Unitless	Relative relief of the basin is divided by the length of the basin	Schumm (1956)	Low Rr is due to the large size and shape of the basin (Gottschalk 1964). Rr indicates the overall steepness of the basin and also indicates denudational process intensity (Schumm 1956). Rr is helpful in inspecting the erosion rate in a basin (Akram et al. 2009). More Rr was observed in hilly areas (Vishwas 2021)
Dissection index (Di)	Di = H/R where H = relative relief (m) and R = absolute relief (m)	Unitless	Relative relief of the basin is divided by the absolute relief of basin	Schumm (1956)	Values range from 0 to 1. Values near 0 indicate less vertical erosion and are not dominant, and a flat terrain is observed, and values near 1 indicate nearly vertical cliff which is a very exceptional case. High Di value was observed in the mountains than in the plain plateau (Mahala 2020)
Ruggedness index (Ri)	Ri = Dd * H/1000 where Dd = drainage density and H = relative relief	Unitless	Ri is the product of the Dd and the relative relief of the basin	Schumm (1956)	Simply indicates the smoothness or roughness of a surface and the proneness of surface to soil erosion. The value of Ri will be high when both relative relief and Dd are long and steep slopes (Strahler 1957). High Ri suggests that geomorphic development is in its early stage, and the surface is rough. The Mountain environment has high Di values

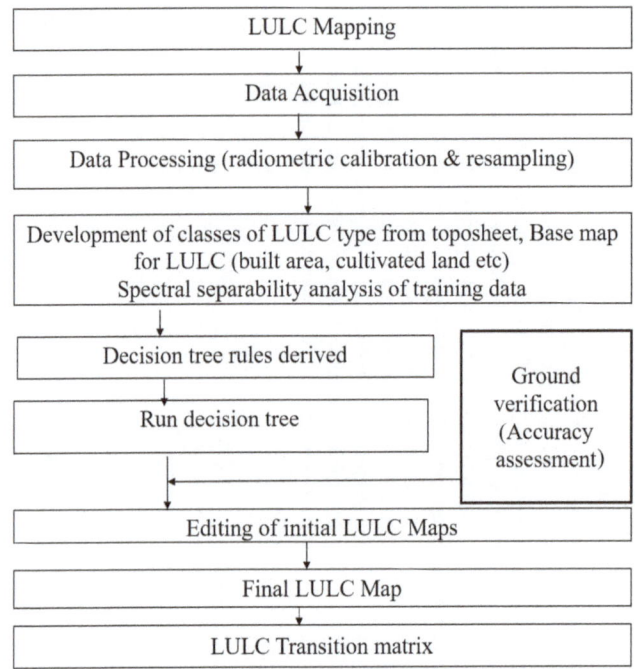

Fig. 1.4 Methodology for LULC mapping of the study area

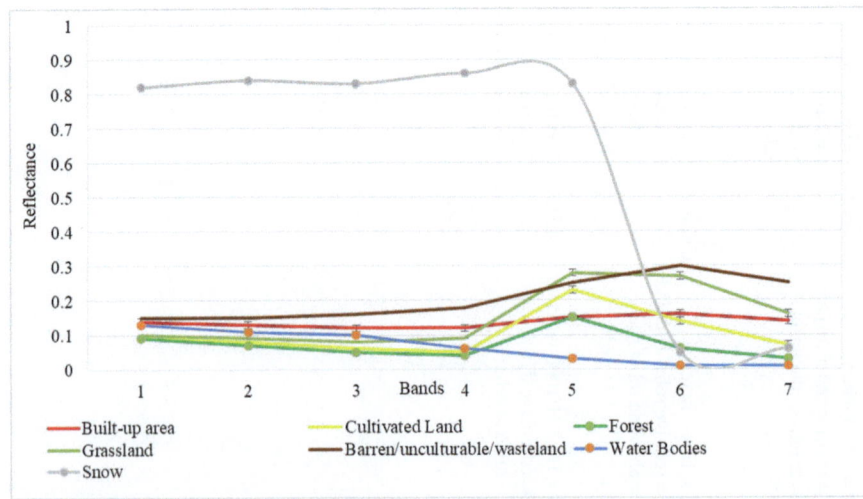

Fig. 1.5 Spectral profile of different LULC classes (based on Landsat 8 data)

Fig. 1.6 LULC categories of the study area: (1) Kullu city; (2) apple and other crops near Sarsai village; (3) deodar forest near Manali city; (4) grassland area near Buruwa village; (5) rocks near Solang valley road; (6) Beas River near Palchan village; (7) snow cover near Dhundi

With the use of toposheets and satellite pictures, training examples for all LULC classes were created. Decision criteria were generated from the training samples' spectral profiles' separability analysis. One of the LULC classes' spectral profiles based on Landsat OLI is shown in Figs. 1.5 and 1.6.

The Landsat MSS and TM profiles were also created similarly. Additionally, from the spectral profiles of training samples for LULC classes, descriptive statistics were calculated. With the use of local knowledge, the band ratios and their threshold that are best for identifying various LULC classes were determined. Some indices, including the normalized difference built-up index (NDBI), the normalized difference vegetation index (NDVI), the water ratio index (WRI), and the normalized difference snow index (NDSI), were also used in the image to enhance the classification result. A decision tree and node expressions were created for LULC categorization of the research region based on Landsat 8 data (Fig. 1.7). Because the threshold values for object reflection and indices fluctuate with the seasons, a decision tree with a similar design was also created for the categorization of the remaining years' LULC data.

The decision tree's nodes' expressions are all displayed in Table 1.5. NDSI was used in Node 1 to categorize the image into snowy conditions and those without snow. Using trial-and-error techniques, an NDSI threshold of 0.4 was decided upon. Next, node 2-1 is based on the WRI, in which water has a threshold value of ≥ 1. Based on NDVI expressions, Node 3-1 divided the remaining image into vegetation and non-vegetation-based threshold values of 0.3. The built-up area and BUW are additional categories for the non-vegetation area.

Shapefiles for cultivated land and built-up areas were created using toposheets and the most recent Google Earth imagery (Fig. 1.8). After that, zonal statistics were calculated by overlaying the DEM with the shapefiles for the built-up area and

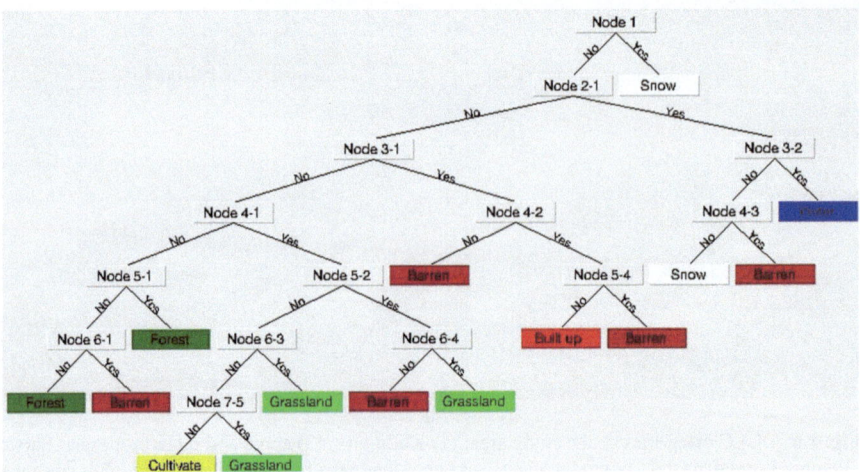

Fig. 1.7 Final decision tree derived from training samples statistics and spectral indices for Landsat 8 data

Table 1.5 Expressions applied at nodes in the decision tree for LULC mapping of the study area

Nodes	Expressions
Node1	((b3 − b6)/(b3 + b6) GT 0.4), (b5 GT 0.09), and (b3 GT 0.10)
Node 2-1	(b3 + b4)/(b5 + b6) GT 1
Node 3-1	((b5 − b4)/(b5 + b4) LT 0.40), ((b7/b3) GE 0.70), and (b7 GT 0.08)
Node 3-2	b10 LT 3000
Node 4-1	((b5 − b4)/(b5 + b4) LT 0.73) and (b6 GT 0.11)
Node 4-2	((b7/b3) LE 1.4) and (b6 LT 0.20)
Node 4-3	b7 GT 0.10
Node 5-1	((b5 − b4)/(b5 + b4) GT 0.75)
Node 5-2	b10 GT 2800
Node 5-4	b10 GT 2700
Node 6-1	b10 GT 3800
Node 6-3	{slope} GT 35
Node 6-4	((b5 − b4)/(b5 + b4) GT 0.50)
Node 7-5	b5 GT 0.30

Note: *GT* greater than, *GE* greater and equal to, *LE* less than and equal to, *LT* less than, *b* band of the satellite, *b10* DEM band

cultivated land (Table 1.6). With the aid of the NDVI, slope, and elevation, the remaining area was divided into forests, grasslands, and cultivated land.

Ground verification was performed for dubious locations using field photography and the Global Positioning System (GPS). The incorrectly categorized regions were changed on the final map based on ground verification. The following variables were estimated to assess the extent and rate of change in the LULC of the study area:

$$C_a = T_{a(t2)} - T_{a(t1)}$$

$$C_e = \frac{C_a}{T_{a(t1)}} \times 100$$

where t_1 and t_2 are the beginning and ending times of the land cover studies conducted.

T_a = total area; C_a = changed area; C_e = change extent

Because various LULC classes have distinct impacts on streamflow, a LULC transition matrix was also created to determine the spatial conversion from one class to another over the analysis period. For instance, the streamflow would be reduced if grassland were turned into forest. In contrast, a rise in snow cover would increase summer streamflow. For the field survey, samples were obtained using simple random selection. The sample of places to be visited was not intended to include locations that were higher than 2400 m or too far from the road system. Validation of

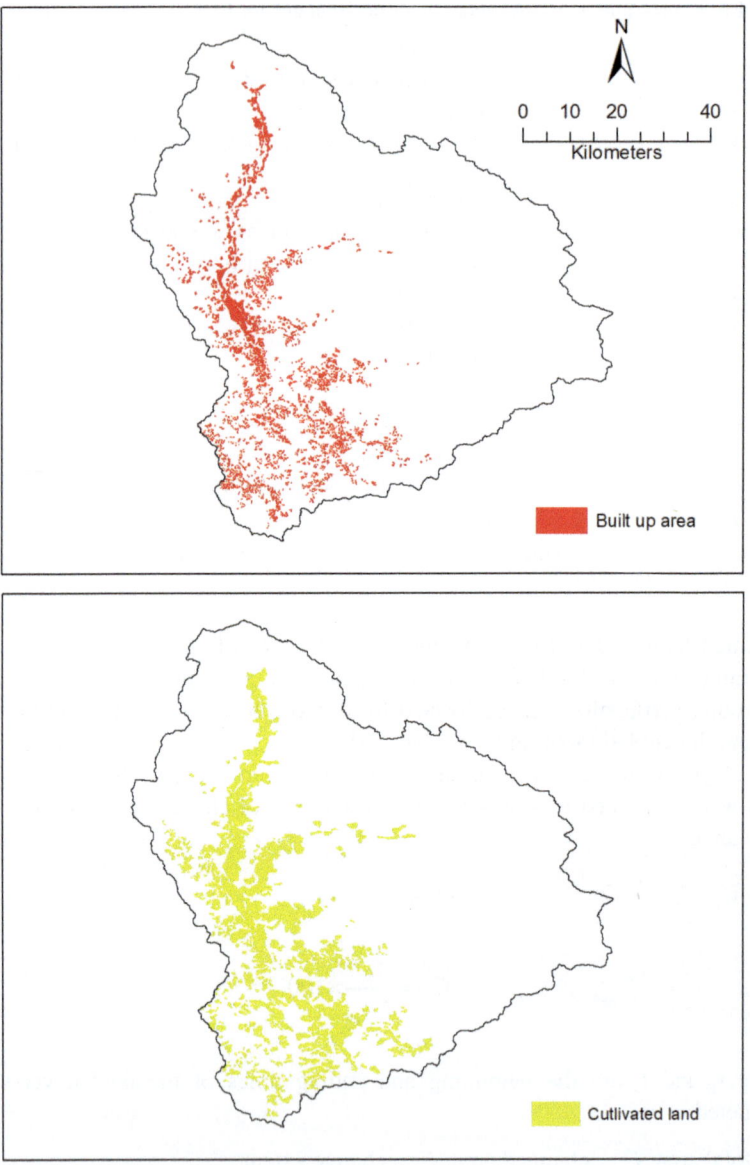

Fig. 1.8 Digitized built-up area and cultivated land in the upper Beas basin of the study area

sample positions over 2400 m was done using either toposheet or Google Earth photos. The field survey was conducted using a Garmin GPS etrex 10. All of the GPS locations recorded during the field survey were captured on camera. A contingency table and error matrix were used to evaluate the precision of the categorized maps. Along with the inclusion (commission errors) and exclusion (omission errors) errors inherent in the classification, the accuracy of each category in the error matrix

Table 1.6 Zonal statistics derived from built-up area, cultivated land, and elevation of the study area

Statistics	Built-up area	Cultivated land
	Elevation (m)	
Minimum	836	861
Maximum	3900	3804
Range	3064	2943
Mean	1811	1816
Standard deviation	664	430

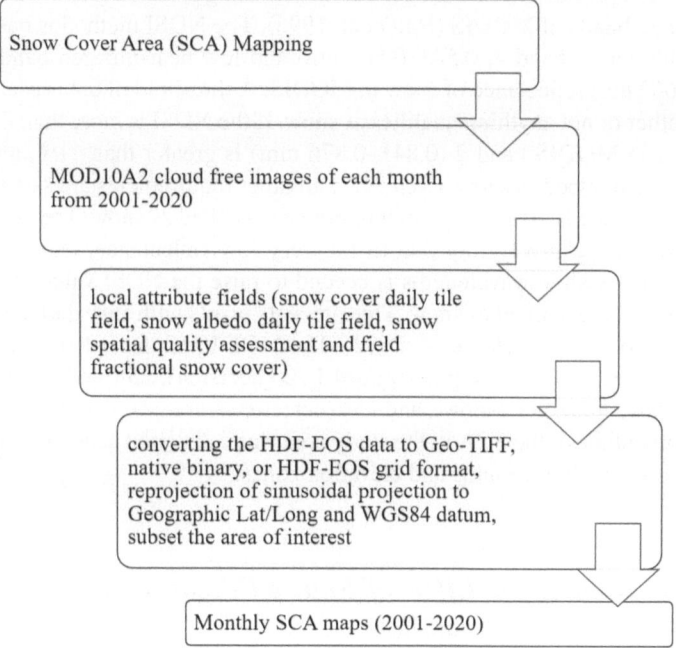

Fig. 1.9 SCA mapping methodology of the present study

is specified (Congalton 1991). Additionally, the categorized map's accuracy was evaluated using the Kappa (K) coefficient (Cohen 1960).

1.9.2.3 Snow Cover Area mapping

The study area's SCA was also mapped using MODIS/Terra Snow Cover Daily L3 Global 500 m Grid (MOD10A2) data (cloud-free images) from January 2001 to December 2020 (Fig. 1.9). Data are gridded in sinusoidal projection using the WGS84 datum, in equal-area tiles. The Advanced Very High-Resolution Radiometer (AVHRR) sensor data from the National Oceanic and Atmospheric Administration

(NOAA) satellite are compared to the MODIS satellite data as having a higher resolution. At 500-m spatial and daily temporal resolutions, MODIS also offers data on
snow (Hall et al. 2002). Positive outcomes have been obtained from investigations
on the accuracy and constraints of the MODIS Terra snow cover products. Under a
clear sky, the daily snow output from MODIS Terra has an overall accuracy of
around 90%. Additionally, MODIS satellite data feature an automatic snow-mapping
technique and a high temporal resolution compared to other satellite data (Landsat
and IRS), which lowers the time and mistakes involved in processing the satellite
data manually.

The NDSI technique is employed to calculate the SCA for the creation of MODIS
snow-covered products. The NDSI is calculated using 2 bands (band 4 and band 6)
out of the 36 bands of MODIS (Hall et al. 1995). The NDSI method is based on the
high visible band (band 4, 0.545–0.565 um) and low near-infrared band (band 6,
1.628–1.652 um) reflectance of snow in MODIS. A threshold of 0.4 is used to determine whether or not anything qualifies as snow. If the NDSI is more than 0.4 and the
reflectance in MODIS band 2 (0.841–0.876 mm) is greater than 11%, a pixel in a
forest will be mapped as snow. Even when all other requirements are satisfied, if the
band 4 reflectance is 10%, the pixel will not be classified as snow. The denominator
in the NDSI is relatively tiny due to the very low reflectance, and just a slight
increase in the visible wavelengths is needed to raise the NDSI value high enough
to mistakenly label a pixel as snow. This prevents pixels with very dark targets from
being mapped as snow (Hall et al. 2002). The MOD10A2 data (daily tile field data
for snow cover) includes the pixel values 1 (no decision), 25 (land without snow),
50 (cloud obscured), 200 (snow), and 254 (detector saturated) (Hall et al. 2007). The
overall percentage of the SCA in the study area was calculated using the categorized
SCA maps for various months and elevation zones.

1.9.3   To Analyze the Effect of Future Climate and LULC Changes on Streamflow

Given that there were only Thalout station discharge data available, the SWAT
model was used to determine how future climatic and LULC changes might affect
streamflow in the upper Beas basin. The watershed's topography, LULC, soil type,
and meteorological data are needed for this model (Fig. 1.10). It is a continuous-
time model that is widely employed to evaluate how climatic variability affects
runoff all across the planet. The software for the current investigation was QGIS and
QSWAT. The coefficient of determination (R^2) and Nash-Sutcliffe efficiency (NSE)
was used to evaluate the model's performance during the calibration and validation
phases (Neitsch et al. 2002; Shi et al. 2011). To calculate their influence on the
basin's streamflow, the SWAT model provided data on the climate. One component
is used at a time, while the other ones are kept constant in this method.

It is expected that the SWAT model is very suited to semiarid climatic conditions.
To adapt the model to the climate in the Himalayas, input parameters for snow

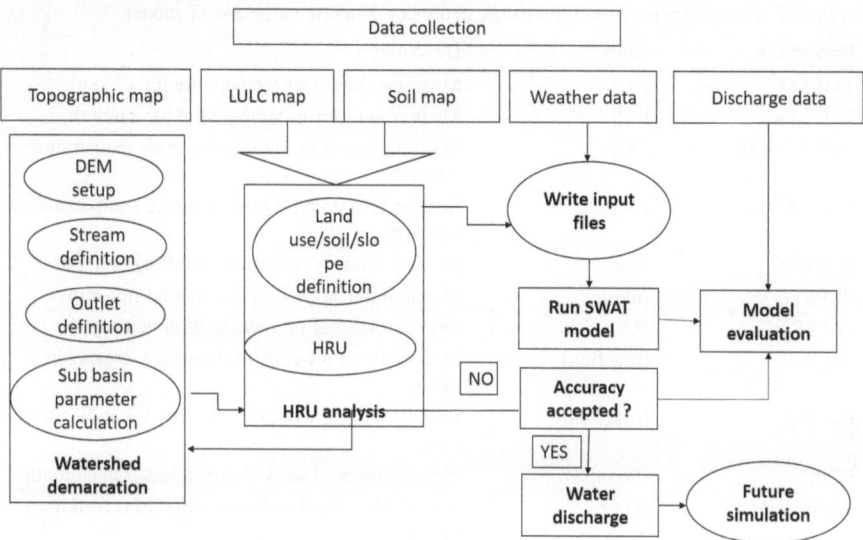

Fig. 1.10 The working flow of the SWAT model

processes were customized depending on basin features. Elevation band, precipitation rate, temperature lapse rate, rain/snow threshold, maximum melt coefficient, minimum melt coefficient, snowpack temperature lag factor, snowpack temperature melt factor, and areal snow coverage threshold CV_{100} and CV_{50} are among these input parameters.

1.9.3.1 SWAT Model Setting

Weather Generator Data

Daily weather information is necessary for the SWAT hydrological model to supply the energy and moisture needed to power all processes simulating in the basin (Neitsch et al. 2011; Arnold et al. 2012). The SWAT model environment's weather model was configured using daily rainfall, minimum temperature, and maximum temperature data during 1969–2020 (Tables 1.7 and 1.8). Following the procedures described in the SWAT model's handbook, these parameters are calculated using meteorological data for the period 1969–2020 to fill up any gaps in weather data during modelling (Neitsch et al. 2002).

Elevation Data

Cartosat-1 satellite data was used to provide elevation information for the study region. By using a Garmin eTrex 10 to capture 500 GPS points, the accuracy of the Cartosat-1 DEM was examined (Fig. 1.11). The calculated R^2 of the ground and the DEM elevation points was 0.99, demonstrating good accuracy (Fig. 1.11). The SWAT model uses DEM for three different things. The first is to create a stream

Table 1.7 Parameters required for weather generator database in the SWAT model

Parameters	Units	Description
TMPMX	[°C]	Mean maximum air temperature for a month
TMPMN	[°C]	Mean minimum air temperature for a month
TMPSTDMX	[°C]	Standard deviation for maximum air temperature in a month
TMPSTDMN	[°C]	Standard deviation for minimum air temperature in a month
PCPMM	[mm]	Mean amount of precipitation falling in a month
PCPSTD	[mm]	Standard deviation for precipitation in a month
PCPSKW	NA	Skew coefficient for precipitation in a month
PR_W1	[Fraction]	Probability of a wet day following a dry day in a month
PR_W2	[Fraction]	Probability of a wet day following a wet day in a month
PCPD	[Days]	Mean number of days of precipitation in a month
RAINHHMX	[mm]	Max 0.5-h rainfall in entire period of record for a month
SOLARAV	[MJ/m^2-day]	Mean daily solar radiation in a month
DEWPT	[°C]	Mean dew point temperature in a month
WNDAV	[m/s]	Mean wind speed in a month

Source: Neitsch et al. (2002)
Note: *NA* not applicable

network to help define the basin. The second is to create a slope to help establish hydrological response units (HRUs), and the third and last step is to develop elevation bands to define the basin.

Land Use/Land Cover Data

The SWAT model used LULC categorized maps created for the basin (Chap. 3). The model's simulation of plant development needs LULC-related parameters (Table 1.9) (Neitsch et al. 2002). Leaf area index (LAI) was used to calculate foliage cover and predict crop growth and production (Boegh et al. 2002; ENVI undated). An LAI map of the study area was created for the years 2000 and 2020 which has values between −1 and 1. Then, zonal statistics for each LULC class were calculated and are shown in Table 1.10 and Fig. 1.12 by superimposing LULC maps on LAI maps of the study area.

Information on plants, including the maximum root depth, the ideal temperature for plant development, and the lowest temperature for plant growth, were obtained from the National Horticulture Board.[3] Using a clinometer, the maximum canopy height of the chosen tree types is calculated during fieldwork at Rangdi (1560 m) and Dhundi (3000 m). A total of 20 tree samples were selected within a 5-by-5-m area (Table 1.11).

---

[3] http://nhb.gov.in/model-project-reports/Horticulture%20Crops/apple/apple1.htm

Table 1.8 Computed weather generator parameters for weather stations

ST	Jan	Feb	Mar	April	May	Jun	Jul	Aug	Sep	Oct	Nov	Dec
TMPMX												
A	10.63	11.79	16.19	21.27	24.67	26.72	25.72	25.26	24.56	22.12	17.95	13.68
B	15.56	17.77	22.20	27.69	31.26	32.80	31.30	30.74	30.05	27.91	22.99	17.77
TMPMN												
A	−0.88	0.13	2.88	6.25	8.86	12.51	15.30	15.23	11.50	5.84	1.89	0.50
B	1.58	3.50	6.58	9.73	12.75	16.71	19.62	19.60	16.36	9.90	4.54	1.63
TMPSTDMX												
A	3.97	4.52	5.47	4.99	4.41	3.66	3.04	2.75	2.75	3.04	3.40	3.70
B	4.04	4.63	5.64	4.63	4.08	3.21	3.18	2.55	3.00	2.93	3.19	3.21
TMPSTDMN												
A	3.65	3.34	2.91	2.66	2.62	3.13	2.77	2.56	3.22	2.47	2.01	3.22
B	2.24	2.35	2.45	2.40	2.68	3.53	2.95	2.63	3.08	2.77	2.20	2.18
PCPMM												
A	110.04	107.57	159.75	100.29	76.34	101.87	194.04	199.62	110.41	46.50	43.36	60.63
B	81.86	78.96	99.93	54.60	53.76	43.71	101.29	95.41	49.22	18.28	18.20	34.82
PCPSTD												
A	9.49	7.90	11.93	8.35	6.50	16.36	10.81	10.75	9.15	5.38	5.33	7.26
B	8.62	7.69	9.75	5.11	5.06	4.79	7.99	8.53	5.22	3.80	3.14	5.22
PCPSKW												
A	7.79	3.30	3.89	4.49	4.69	23.56	3.15	3.13	7.01	9.20	7.71	7.40
B	6.55	4.59	8.67	4.41	5.13	6.61	4.47	7.34	6.03	15.94	8.90	10.23
PR_W1												
A	0.15	0.18	0.19	0.15	0.17	0.24	0.34	0.40	0.20	0.09	0.06	0.08
B	0.13	0.20	0.21	0.19	0.23	0.22	0.32	0.28	0.16	0.07	0.04	0.08

(continued)

Table 1.8 (continued)

ST	Jan	Feb	Mar	April	May	Jun	Jul	Aug	Sep	Oct	Nov	Dec
PR_W2												
A	0.72	0.69	0.67	0.71	0.57	0.58	0.70	0.71	0.66	0.69	0.73	0.67
B	0.61	0.54	0.54	0.51	0.42	0.47	0.62	0.63	0.55	0.41	0.67	0.64
PCPD												
A	11.45	11.33	11.98	11.17	9.12	11.52	17.10	18.55	11.95	7.31	6.14	6.40
B	8.12	9.24	10.24	8.62	9.31	9.31	14.55	13.76	8.48	3.48	3.71	5.93
RAINHHMX												
A	60.00	19.80	37.47	25.40	21.53	167.33	33.33	29.43	50.93	29.33	26.53	35.33
B	33.33	24.20	66.67	17.80	18.03	20.87	26.43	44.00	24.30	30.73	14.33	32.93
SOLARAV												
A	13.91	16.04	17.95	18.70	17.00	15.52	13.65	14.06	16.45	16.97	15.27	13.30
B	14.35	16.77	18.50	19.41	17.55	15.39	13.41	14.21	16.67	17.65	15.83	13.72
DEWPT												
A	0.50	−0.71	1.16	2.32	3.64	7.39	12.57	12.93	10.23	4.77	2.16	1.23
B	4.40	4.82	6.03	6.92	8.11	11.57	17.00	17.58	14.86	9.24	6.19	5.26
WNDAV												
A	0.55	0.79	0.94	1.05	0.97	1.13	1.20	1.22	1.14	0.94	0.65	0.42
B	0.55	0.79	0.94	1.05	0.97	1.13	1.20	1.22	1.14	0.94	0.65	0.42

Note: Grid location A-Y321X771, B- Y315X771

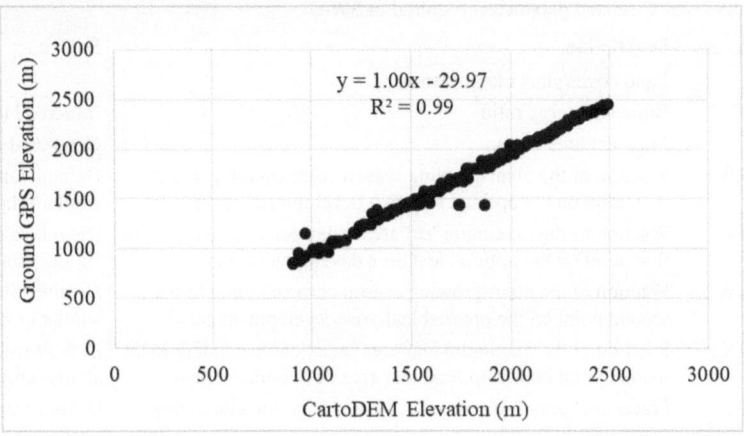

Fig. 1.11 Accuracy of the DEM of the study area

Soil Data

The obtained soil's physical and chemical properties were used to calculate SWAT's soil parameters (soil hydrologic group, moist bulk density, an available water capacity of the soil layer, and saturated hydraulic conductivity) using the Soil Water Characteristics software developed by Saxton (1986, 2006) (Table 1.12).

The soil erodibility (K) factor is computed by Williams (1995) (Neitsch et al., 2002) using the Universal Soil Loss Equation (USLE). According to Brady's (1984) formula, organic matter (OM) is calculated as follows:

$$OM = 1.72 \times OC (\text{organic carbon in} \%)$$

Based on the soil color code mentioned in the report of SLUSI (2013), moist soil albedo is calculated. Following the computation of their damp soil albedo, their values were taken from the Munsell[4] color based on the kind of color:

$$\text{Moist soil albedo} = 0.069 \times (\text{color values}) - 0.114$$

1.9.3.2   SWAT Model Setup

Watershed Delineation

The upper Beas basin and sub-basins were delineated from the DEM (30 m) using the threshold size of 50 km². A total of 19 sub-basins were created in the upper Beas basin (Fig. 1.13).

---

[4] https://nenc.gov.ua/old//GLOBE/Other/Munsell%20soil%20colour%20chart.pdf

Table 1.9 LULC-related parameters required in SWAT

Parameters	Description	Source
IDC	Land cover/plant classification	
BIO_E	Biomass/energy ratio	Infocrop2 model
BLAI	Max leaf area index	Computed
FRGRW1	Fraction of the plant-growing season corresponding to the first point on the optimal leaf area development curve	Default value of similar crop
LAIMX1	Fraction of the maximum leaf area index corresponding to the first point on the optimal leaf area development curve	Default value of similar crop
FRGRW2	Fraction of the plant-growing season corresponding to the second point on the optimal leaf area development curve	Default value of similar crop
LAIMX2	Fraction of the maximum leaf area index corresponding to the second point on the optimal leaf area development curve	Default value of similar crop
DLAI	Fraction of growing season when leaf area starts declining	Default value of similar crop
CHTMX	Max canopy height	Computed
RDMX	Max root depth	Computed
T_OPT	Optimal temp for plant growth	Computed
T_BASE	Mini temp plant growth	Computed
Cropname	Crop description name	Scientific name
ALAI_MIN	Minimum leaf area index for plant during dormant period	Computed
EXT_ COEF	Light extinction coefficient	Infocrop2 model
BM_ DIEOFF	Biomass die-off fraction	Default value of similar crop

Source: Neitsch et al. (2011)

Table 1.10 Estimated leaf area index related parameter for the SWAT model

LULC classes	BLAI	LAIMX1	LAIMX2	DLAI	ALAI_MIN
Apple	4	0.15	0.75	0.55	0.75
Forest	5.00	0.70	0.99	0.60	0.75
Grassland	2.50	0.10	0.70	0.35	0
Barren/unculturable/wasteland (BUW)	0.01	0.05	0.06	0.01	0
Water bodies	0	0	0	0	0
Snow	0	0	0	0	0

Note: *BLAI* maximum leaf area index, *LAIMX1* fraction of the maximum leaf area index corresponding to the first point on the optimal leaf area development curve, *LAIMX2* fraction of the maximum leaf area index corresponding to the second point on the optimal leaf area development curve, *DLAI* fraction of growing season when leaf area starts declining, *ALAI_MIN* minimum leaf area index for plant during dormant period

Hydrological Response Units

In the current study, 180 hydrological response units (HRUs) were produced using land use, soil, and slope (Fig. 1.13). These discontinuous areas of a sub-basin have distinctive characteristics in terms of land use, management, and soil, and grouping similar soil and land use areas into a response unit make running the model easier.

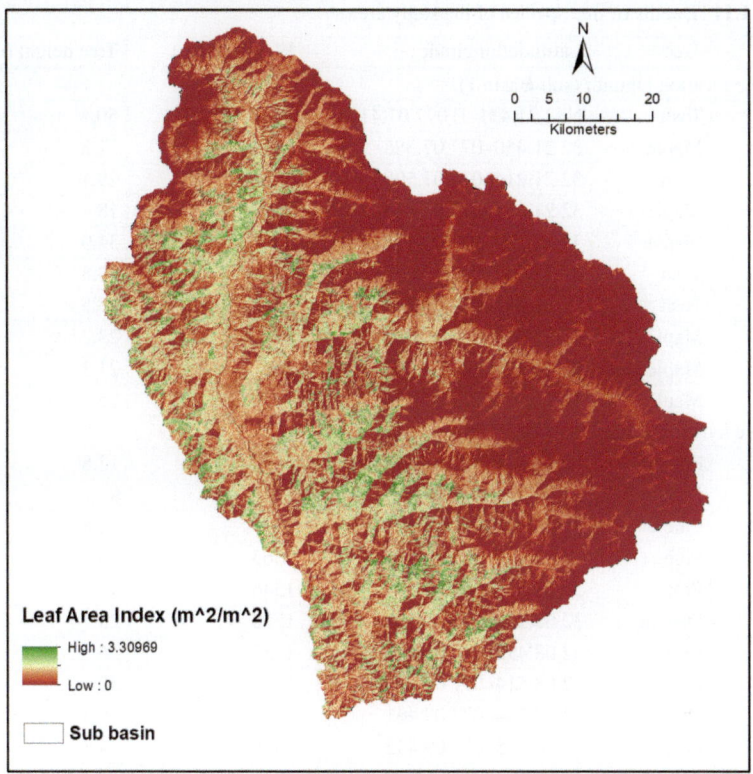

Fig. 1.12 Distribution of estimated leaf area index of the study area, 2000

Four types of slopes were established: 10%, 10–15%, 15–25%, and >25% (Table 1.13). The majority of the basin has a slope of more than 85%, suggesting a steep to an extremely steep slope.

1.9.3.3 Elevation Band

The elevation band is crucial for a basin that gets precipitation in the form of both rain and snow. The SWAT model provides for up to ten elevation bands to account for orographic impacts on both rainfall and air temperature. Elevation bands brought about constant variations in water distribution across the watersheds' hydrological cycle (Grusson et al. 2015). As a consequence of the relative lapse rates, the model determines rainfall as well as the maximum and minimum temperature for each band of elevation. The number of elevation bands was chosen based on zonal data, which were calculated by using the monthly map of the basin's maximum and minimum SCA for the period 2001–2020. The different elevation zones are used to calculate the SCA (Tables 1.14 and 1.15). The 850-class interval of elevation band was chosen for the SWAT model based on SCA under various elevation zones since it

Table 1.11 Details of tree species of the study area

S. no	Tree	Latitude/longitude	Elevation (m)	Tree height (m)
Sample location Dhundi (sub-Basin 1)				
1	Tosh	N32 21.451–E 077 07.419	3017	50.4
2	Maple	32 21.450–077 07.388	3011	31.8
3	Tosh	32 21.214–077 07.508	2850	29.1
4	Maple	32 21.374–077 07.483	2927	18
5	Maple	32 21.109–077 07.630	2889	31.9
6	Tosh	32 21.095–077 07.662	2866	23.8
7	Tosh	32 21.057–077 07.788	2814	23.5
8	Maple	32 21.410–077 07.484	2949	23.2
9	Maple	32 21.416–077 07.422	2957	21.3
10	Maple	32 21.398–077 07.450	2959	25.3
Sample location Rangri (sub-Basin 39)				
11	Poplar	32 08.414–077 09.466	1537	12.8
12	Poplar	32 08.422–077 09.444	1536	9
13	Pine	32 08.446–077 09.507	1549	34.8
14	Pine	32 08.488–077 09.526	1563	20.8
15	Pine	32 08.514–077 09.530	1546	45.2
16	Deodar	32 08.542–077 09.541	1558	28
17	Deodar	32 08.492–077 09.496	1557	36
18	Deodar	32 08.514–077 09.464	1552	30
19	Deodar	32 08.534–077 09.461	1556	26.3
20	Pine	32 08.565–077 09.452	1589	34.8

Source: Field Work 2017

Table 1.12 Soil-related parameters required for the model

Parameters	Units	Description	
NLAYERS	NA	Number of layers in the soil	Report[a]
HYDGRP	NA	Soil hydrologic group	Computed
SOL_ZMX	[mm]	Maximum rooting depth of soil profile	Report[a]
Texture	NA	Texture of soil layer	Report[a]
SOL_Z	[mm]	Depth from soil surface to bottom of layer	Report[a]
SOL_BD	[g/cm^3]	Moist bulk density	Computed
SOL_AWC	[mm/mm]	Available water capacity of the soil layer	Computed
SOL_K	[mm/h]	Saturated hydraulic conductivity	Computed
SOL_CBN	[%]	Organic carbon content	Report[a]
Clay	[%]	Clay content	Report[a]
Silt	[%]	Silt content	Report[a]
Sand	[%]	Sand content	Report[a]
Rock	[%]	Rock fragment content	Report[a]
SOL_ALB	NA	Moist soil albedo	Computed
USLE_K	NA	USLE equation soil erodibility (K) factor	Computed

Note: [a]SLUSI, 2013; *NA* not applicable
Source: Neitsch et al. (2002)

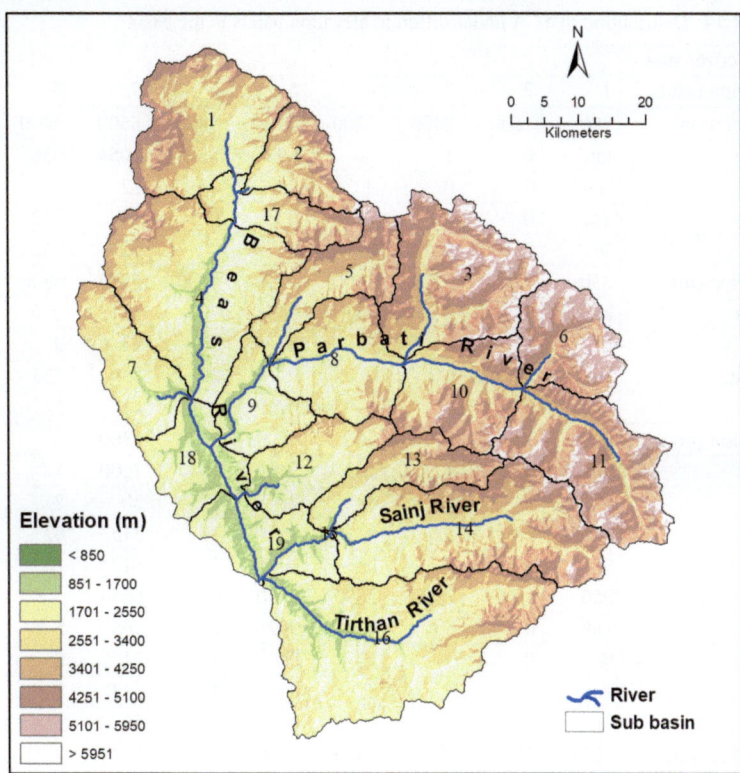

Fig. 1.13 Extent and river system of the upper Beas basin up to Thalout and its sub-basins

Table 1.13 Slope distribution in the upper Beas basin

Slope (%)	% of total basin area
< 10	7.16
10–15	3.2
15–25	6.93
> 25	82.71

was thought to be the basin's best representation of SCA (Fig. 1.13). A maximum of around 24% of the basin is covered by elevation band 3, which is followed by bands 4 and 6 (Fig. 1.13 and Table 1.14).

Following the development of HRUs for the basin, default values were placed in the model's input files for weather, land use, soil, groundwater, management, etc. The SWAT model database tables, including parameter values, were then all put into the project database for the basin. The first 5 years of data were used to warm up the model and establish the basin's basic hydrological conditions. The model was configured for 19 sub-basins for the years 1974–2020, ignoring the first 5 years

Table 1.14 Distribution of SCA under different elevation bands in the basin

Snow cover area									
Elevation bands	1	2	3	4	5	6	7	8
Elevation (m)	800	1600	2400	3200	4000	4800	5600	6400
March	km²	0	1	86	697	824	1054	636	12
	%	0	0	3	21	25	32	19	0
August	km²	0	0	0	0	0	47	322	17
	%	0	0	0	0	0	12	83	4
Elevation (m)	850	1700	2550	3400	4250	5100	5950	6800
March	km²	0	2	153	847	906	1112	290	1
	%	0	0	5	26	27	34	9	0
August	km²	0	0	0	0	2	159	223	2
	%	0	0	0	0	1	41	58	0
Elevation (m)	900	1800	2700	3600	4500	5400	6300	
March	km²	0	4	263	940	1047	1000	57	
	%	0	0	8	28	32	30	2	
August	km²	0	0	0	0	10	313	62	
	%	0	0	0	0	3	81	16	
Elevation (m)	950	1900	2850	3800	4750	5700	6650	
March	km²	0	6	404	997	1190	708	5	
	%	0	0	12	30	36	21	0	
August	km²	0	0	0	0	34	343	8	
	%	0	0	0	0	9	89	2	
Elevation (m)	1000	2000	3000	4000	5000	6000	7000	
March	km²	0	11	558	1040	1302	399	1	
	%	0	0	17	31	39	12	0	
August	km²	0	0	0	0	113	272	1	
	%	0	0	0	0	29	70	0	

Table 1.15 Elevation bands division of the upper Beas River basin

Elevation bands	Elevation (m)	% of total basin area
1	<850	0.02
2	850–1700	9.10
3	1700–2550	23.80
4	2550–3400	22.91
5	3400–4250	16.87
6	4250–5100	21.23
7	5100–5950	6.05
8	> 5900	0.03

(1969–1973). But for calibration and validation, just one sub-basin with observed flow records was used. When conducting the model, the skewed normal rainfall distribution and the Hargreaves ET technique were taken into account. The surface flow in the basin was modelled using the Soil Conservation Service (SCS) curve number approach (Neitsch et al. 2011). One element (climate or land cover) from the baseline period was kept constant at a time when the SWAT model was applied using daily weather produced using various climate and land cover change scenarios for the period 2020–2071. The model shows how the projected flow will differ from the baseline (1974–2020) by 2071. According to official records, 4.8% of the total land is irrigated for crops during 2011–2012.[5] Surface water is the primary source of irrigation in the basin, although statistics on the volume of water used for irrigation is not available. So the model does not include irrigation data. The upper Beas basin is home to both micro- and major hydropower projects. Due to a lack of available discharge data from the operating projects, the analysis did not account for their impact on streamflow in the basin by 2071.

1.10 SWAT Model Result Analysis

The Mann-Kendall test (Mann 1945; Kendall 1975) and Theil Sen Slope estimator (Theil 1950; Sen 1968) non-parametric tests were used to evaluate the trend and rate of trend in the mean monthly flow at Thalout for the period 1971–2002. For the period 2020–2071, the effects of changing climatic variables (such as air temperature and rainfall) and land cover have been assessed in terms of the resultant streamflow at monthly, seasonal, and annual scales. The study compares the baseline period (1974–2020) water availability to the future period after first determining the baseline period's water availability (2020–2071). The following formula was used to calculate the expected mean flow percentage change from the baseline era (1974–2020) to the late century (2071).

$$\frac{\text{predicted period mean monthly flow} - \text{reference period mean monthly flow}}{\text{reference period mean monthly flow}} \times 100$$

1.11 Scope of the Book

There are seven chapters in this book. The background, issue statement, data sources, and research approach are all covered in the first chapter. The physical and demographic characteristics of the study area were covered in the second chapter. The third chapter discusses the basin's changing climate and its variation. The

---

[5] http://aps.dac.gov.in/LUS/Public/Reports.aspxs

fourth chapter explained the LULC changes in the area of research. The fifth chapter demonstrates how changes in the LULC and climate have affected the basin's hydrology. The sixth chapter discusses the many programs and policies implemented by the national and state governments to address the climatic and LULC changes in the study region. A summary and conclusion of the book are provided in the last chapter.

References

Akhter G, Ge Y, Iqbal N, Shang Y, Hasan M (2021) Appraisal of remote sensing technology for groundwater resource management perspective in Indus Basin. Sustainability 13(17):9686

Akram J, Khanday MY, Rizwan A (2009) Prioritization of subwatersheds based on morphometric and land use analysis using remote sensing and GIS techniques. J Indian Soc Remote Sens 37(2):261–274. https://doi.org/10.1007/s12524-009-0016-8

Ali H, Modi P, Mishra V (2019a) Increased flood risk in Indian sub-continent under the warming climate. Weather Clim Extreme 25:100212. https://doi.org/10.1016/j.wace.2019.100212

Ali SH, Shafqat MN, Eqani SA, Shah ST (2019b) Trends of climate change in the upper Indus basin region, Pakistan: implications for cryosphere. Environ Monit Assess 191(2):1–2

Ali S, Cheema MJ, Waqas MM, Waseem M, Awan UK, Khaliq T (2020) Changes in snow cover dynamics over the Indus Basin: evidences from 2008 to 2018 MODIS NDSI trends analysis. Remote Sens 12(17):2782

Allan RP, Slingo A, Ramaswamy V (2002) Analysis of moisture variability in the European Centre for Medium-Range Weather Forecasts 15-year reanalysis over the tropical oceans. J Geophys Res Atmos 107(D15):ACL-1

Anand J, Gosain AK, Khosa R (2018) Prediction of land use changes based on Land Change Modeler and attribution of changes in the water balance of Ganga basin to land use change using the SWAT model. Sci Total Environ 644:503–519. https://doi.org/10.1016/J.SCITOTENV.2018.07.017

Archer DR, Forsythe N, Fowler HJ, Shah SM (2010) Sustainability of water resources management in the Indus Basin under changing climatic and socio economic conditions. Hydrol Earth Syst Sci 14(8):1669–1680

Arnold JG, Srinivasan R, Muttiah RS, Williams JR (1998) Large area hydrologic modeling and assessment part I: model development. J Am Water Resour Assoc 34(1):73–89

Arnold JG, Kiniry JR, Srinivasan R, Williams JR, Haney EB, Neitsch SL (2012) Soil and water assessment tool-input/ouput documentation, version 2012. Available via http://swat.tamu.edu/documentation/. Accessed 20 Mar 2018

Ashraf A, Rehman H (2019) Upstream and downstream response of water resource regimes to climate change in the Indus River basin. Arab J Geosci 12(16):1

Azam MF, Kargel JS, Shea JM, Nepal S, Haritashya UK, Srivastava S, Maussion F, Qazi N, Chevallier P, Dimri AP, Kulkarni AV (2021) Glaciohydrology of the Himalaya-Karakoram. Science 373(6557):eabf3668

Baig S, Sayama T, Yamada M. (2021) Impacts of climate change on river flows in the upper Indus basin and its subbasins. https://doi.org/10.21203/rs.3.rs-404691/v1

Bajracharya AR et al (2018) Climate change impact assessment on the hydrological regime of the Kaligandaki Basin, Nepal. Sci Total Environ 625:837–848. https://doi.org/10.1016/j.scitotenv.2017.12.332

Behrendt H, Opitz D (2000) Retention of nutrients in river systems: dependence on specific runoff and hydraulic load. Hydrobiologia 410:111–122

Bevacqua E, Zappa G, Lehner F, Zscheischler J (2022) Precipitation trends determine future occurrences of compound hot–dry events. Nat Clim Chang 12(4):350–355

Bhakra Beas Management Board (BBMB) (1988) Snow hydrology studies in India with particular reference to the Satluj and Beas catchments. In: Proceeding of workshop on snow hydrology, Manali, pp 1–14

Bhatta B, Shrestha S, Shrestha PK, Talchabhadel R (2019) Evaluation and application of a SWAT model to assess the climate change impact on the hydrology of the Himalayan River Basin. Catena 181:104082

Bhutiyani M, Kale V, Pawar N (2008) Changing streamflow patterns in the rivers of northwestern Himalaya: implications of global warming in the 20th century. Curr Sci 9(5):618–626

Bhutiyani M, Kale V, Pawar N (2010) Climate change and the precipitation variations in the northwestern Himalaya: 1866–2006. Int J Climatol 30:535–548

Bilal H, Chamhuri S, Mokhtar MB, Kanniah KD (2019) Recent snow cover variation in the upper Indus basin of Gilgit Baltistan, Hindukush Karakoram Himalaya. J Mt Sci 16(2):296–308

Bilal H, Siwar C, Bin Mokhtar M, Kanniah KD, Govindan R, Al-Ansari T (2021) Assessment of land use and land cover changes in association with hydrometeorological parameters in the Upper Indus Basin. J Himal Earth Sci 54(2)

Boegh E, Soegaard H, Broge N, Hasager CB, Jensen NO, Schelde K, Thomsen A (2002) Airborne multispectral data for quantifying leaf area index, nitrogen concentration, and photosynthetic efficiency in agriculture. Remote Sens Environ 81(2–3):179–193

Bøggild CE, Knudby CJ, Knudsen MB, Starzer W (1999) Snowmelt and runoff modelling of an Arctic hydrological basin in west Greenland. Hydrol Process 13(12–13):1989–2002

Borah DK, Bera M, Xia R (2004) Storm event flow and sediment simulations in agricultural watersheds using DWSM. Trans ASABE 47(5):1539

Borgaonkar HP, Pant GB, Rupa KK (1994) Dendroclimatic reconstruction of summer precipitation at Srinagar, Kashmir, India since the late 18th century. The Holocene 4(3):299–306

Borgaonkar HP, Pant GB, Rupa KK (1996) Ring-width variations in Cedrus deodara and its climatic response over the western Himalaya. Int J Climatol 16(12):1409–1422

Brady NC (1984) The nature and properties of soil. Macmillan Book Co., New York

Cannon F, Carvalho LMV, Jones C, Bookhagen B (2015) Multi-annual variations in winter westerly disturbance activity affecting the Himalaya. Clim Dyn 44:441–455. https://doi.org/10.1007/s00382-014-2248-8

Cartosat 1 Data User's Handbook. Available via http://www.euromap.de/download/P5_data_user_handbook.pdf. Accessed 20 Dec 2019

Cartosat-1 DEM. Available via http://bhuvannuis.nrsc.gov.in/bhuvan/web/. Accessed 20 Dec 2019

Chen B, Liu Z (2016) Global water vapor variability and trend from the latest 36 year (1979 to 2014) data of ECMWF and NCEP reanalyses, radiosonde, GPS, and microwave satellite. J Geophys Res Atmos 121(19):11–442

Chorley RJ (2019) Introduction to physical hydrology. Routledge, London

Cohen J (1960) A coefficient of agreement for nominal scales. Educ Psychol Meas 20(1):37–46

Compagno L, Huss M, Zekollari H, Miles ES, Farinotti D (2022) Future growth and decline of high mountain Asia's ice-dammed lakes and associated risk. Commun Earth Environ 3(1):1–9

Congalton RG (1991) A review of assessing the accuracy of classifications of remotely sensed data. Remote Sens Environ 37:35–46

Copernicus Climate Change Service (C3S) (2017) ERA5: fifth generation of ECMWF atmospheric reanalyses of the global climate. Copernicus Climate Change Service Climate Data Store (CDS). https://cds.climate.copernicus.eu/cdsapp#!/home

Corbane C, Florczyk AJ, Pesaresi M, Politis P, Syrris V (2018) GHS built-up grid, derived from Landsat, multitemporal (1975-1990-2000-2014), R2018A. European Commission, Joint Research Centre (JRC). https://doi.org/10.2905/jrc-ghsl-10007. PID: Available via http://data.europa.eu/89h/jrc-ghsl-10007

Corbane C, Pesaresi M, Kemper T, Politis P, Florczyk AJ, Syrris V, Melchiorri M, Sabo F, Soille P (2019) Automated global delineation of human settlements from 40 years of Landsat satellite data archives. Big Earth Data 3(2):140–169. https://doi.org/10.1080/20964471.2019.1625528

Dahri ZH, Ludwig F, Moors E, Ahmad S, Ahmad B, Ahmad S, Riaz M, Kabat P (2021) Climate change and hydrological regime of the high-altitude Indus basin under extreme climate scenarios. Sci Total Environ 768:144467

Dame J, Schmidt S, Müller J, Nüsser M (2019) Urbanisation and socio-ecological challenges in high mountain towns: insights from Leh (Ladakh), India. Landsc Urban Plan 189:189–199

Dimri AP, Dash SK (2012) Wintertime climatic trends in the western Himalayas. Clim Chang 111(3):775–800

Dimri AP, Kumar D, Choudhary A, Maharana P (2019) Future changes over the Himalayas: maximum and minimum temperature. Glob Planet Chang 162:212–234. https://doi.org/10.1016/J.GLOPLACHA.2018.01.015

Dutta S, Ramanathan AL, Linda A (2012) Glacier fluctuation using satellite data in Beas basin, 1972–2006, Himachal Pradesh, India. J Earth Syst Sci 121(5):1105–1112

Dwarakish GS, Ganasri BP (2015) Impact of land use change on hydrological systems: a review of current modeling approaches. Cogent Geosci 1(1):1115691. https://doi.org/10.1080/2331204 1.2015.1115691

Engelhardt M, Leclercq P, Eidhammer T, Kumar P, Landgren O, Rasmussen R (2017) Meltwater runoff in a changing climate (1951–2099) at Chhota Shigri Glacier, Western Himalaya Northern India. Ann Glaciol 58:47–58. https://doi.org/10.1017/aog.2017.13

ENVI Services Engine 5.3 User Guide (undated) Available via https://www.harrisgeospatial.com/docs/pdf/ESE_help.pdf. Accessed 20 Mar 2016

Fayaz A, Singh H, Ahmed P (2020) Assessment of spatiotemporal changes in land use/land cover of North Kashmir Himalayas from 1992 to 2018. Model Earth Syst Environ 6(2):1189–1200

Florczyk AJ, Corbane C, Ehrlich D, Freire S, Kemper T, Maffenini L, Melchiorri M, Pesaresi M, Politis P, Schiavina M, Sabo F, Zanchetta L (2019) GHSL data package, EUR 29788 EN. Publications Office of the European Union, Luxembourg, 2019. ISBN 978-92-76-13186-1. https://doi.org/10.2760/290498. JRC 117104 GHS_SMOD_POP2015_GLOBE_R2019A_54009_1K_V2_0_25_5 and GHS_BUILT_LDSMT_GLOBE_R2018A_3857_30_V2_0_19_9

Forsythe N, Fowler HJ, Blenkinsop S, Burton A, Kilsby CG, Archer DR, Harpham C, Hashmi MZ (2014) Application of a stochastic weather generator to assess climate change impacts in a semi-arid climate: the Upper Indus Basin. J Hydrol 517:1019–1034

Forsythe N, Fowler HJ, Li X-F, Blenkinsop S, Pritchard D (2017) Karakoram temperature and glacial melt driven by regional atmospheric circulation variability. Nat Clim Chang 7(9). https://doi.org/10.1038/nclimate3361

Freire S, MacManus K, Pesaresi M, Doxsey-Whitfield E, Mills J. (2016) Development of new open and free multi-temporal global population grids at 250 m resolution. Population 250

Frey H, Paul F, Strozzi T (2012) Compilation of a glacier inventory for the western Himalayas from satellite data: methods, challenges, and results. Remote Sens Environ 124:832–843

Friedl M, Sulla-Menashe D (2019) MCD12Q1 MODIS/Terra+Aqua Land Cover Type Yearly L3 Global 500m SIN Grid V006 [MCD12Q1.A2001001.h24v05.006.2018142223956; MCD12Q1.A2020001.h24v05.006.2021360060254]. NASA EOSDIS Land Processes DAAC. Available via https://doi.org/10.5067/MODIS/MCD12Q1.006. Accessed 20 Mar 2021

Garg V, Nikam BR, Thakur PK, Aggarwal SP, Gupta PK, Srivastav SK (2019) Human-induced land use land cover change and its impact on hydrology. HydroResearch 1:48–56. https://doi.org/10.1016/j.hydres.2019.06.001

González-Zeas D, Erazo B, Lloret P, De Bièvre B, Steinschneider S, Dangles O (2019) Linking global climate change to local water availability: limitations and prospects for a tropical mountain watershed. Sci Total Environ 650:2577–2586. https://doi.org/10.1016/j.scitotenv.2018.09.309

Gottschalk LC. (1964) Reservoir sedimentation. In: Handbook of applied hydrology. McGraw Hill Book Company, New York, Section. 7

Government of Himachal Pradesh (2002a) Himachal Pradesh human development report. Available via from www.in.undp.org/.../human_develop_report_himachal_pradesh_2002_full_report. pdf? Accessed 20 May 2014

Government of Himachal Pradesh (2002b) State of the environment report on Himachal Pradesh. Department of Environment, Science and technology, Shimla. Available via http://envfor.nic. in/sites/default/files/SoER%20Himachal%20Pradesh_0.pdf. Accessed 30 May 2014

Government of Himachal Pradesh (2012) State strategy and action plan on climate change on Himachal Pradesh. Department of Environment, Science and Technology, Shimla. Available via http://www.moef.nic.in/sites/default/files/sapcc/Himachal-Pradesh.pdf. Accessed 30 May 2014

Grusson Y, Sun X, Gascoin S, Sauvage S, Raghavan S, Anctil F, Sáchez-Pérez JM (2015) Assessing the capability of the SWAT model to simulate snow, snow melt and streamflow dynamics over an alpine watershed. J Hydrol 531:574–588

Guhathakurta P, Rajeevan M (2008) Trends in the rainfall pattern over India. Int J Climatol 28(11):1453–1469

Gupta A, Dimri AP, Thayyen R, Jain S, Jain S (2020) Meteorological trends over Satluj River Basin in Indian Himalaya under climate change scenarios. J Earth Syst Sci 129:161. https://doi. org/10.1007/s12040-020-01424-x

Haleem K, Khan AU, Ahmad S, Khan M, Khan FA, Khan W, Khan J (2022) Hydrological impacts of climate and land-use change on flow regime variations in upper Indus basin. J Water Clim Chang 13(2):758–770

Hall DK, Riggs GA, Salomonson VV (1995) Development of methods for mapping global snow cover using moderate resolution imaging spectroradiometer data. Remote Sens Environ 54(2):127–140

Hall DK, Riggs GA, Salomonson VV, DiGirolamo NE, Bayr KJ (2002) MODIS snow-cover products. Remote Sens Environ 83(1–2):181–194

Hall D, Riggs G, Salomonson V (2007) Updated daily MODIS/Aqua snow cover daily L3 global 500 m grid. National Snow and Ice Data Center, Digital Media, Boulder

Harlin J (1991) Development of a process oriented calibration scheme for the HBV hydrological model. Nord Hydrol 22(1):15–36

Hassan M, Du P, Mahmood R, Jia S, Iqbal W (2019) Streamflow response to projected climate changes in the Northwestern Upper Indus Basin based on regional climate model (RegCM4. 3) simulation. J Hydro-Environ 27:32–49

Hersbach H, Bell B, Berrisford P, Biavati G, Horányi A, Muñoz Sabater J, Nicolas J, Peubey C, Radu R, Rozum I, Schepers D, Simmons A, Soci C, Dee D, Thépaut J-N (2018) ERA5 hourly data on pressure levels from 1979 to present. Copernicus Climate Change Service (C3S) Climate Data Store (CDS). https://doi.org/10.24381/cds.bd0915c6. Accessed 1 May 2020

Horton RE (1932) Drainage-basin characteristics. Trans Am Geophys Union 13(1):350–361. https://doi.org/10.1029/TR013i001p00350

Horton RE (1945) Erosional development of streams and their drainage basins; hydrophysical approach to quantitative morphology. Geol Soc Am Bull 56(3):275–370. https://doi.org/10.113 0/0016-7606(1945)56%5b275:edosat%5d2.0.co;2

Hughes MK (2001) An improved reconstruction of summer temperature at Srinagar, Kashmir since 1660 AD based on tree-ring width and maximum latewood density of Abies pindrow (Royle) Spach. Palaeobotanist 50:13–19

Huss M, Hock R (2018) Global-scale hydrological response to future glacier mass loss. Nat Clim Chang 8(2):135–140

Hussain D, Kao HC, Khan AA, Lan WH, Imani M, Lee CM, Kuo CY (2020) Spatial and temporal variations of terrestrial water storage in upper Indus basin using GRACE and altimetry data. IEEE Access 8:65327–65339

Hussain A, Cao J, Hussain I, Begum S, Akhtar M, Wu X, Guan Y, Zhou J (2021) Observed trends and variability of temperature and precipitation and their global teleconnections in the upper Indus basin, Hindukush-Karakoram-Himalaya. Atmosphere 12(8):973

IMD (India Meteorological Department) (2016) Annual climate summary. Available via http://www.imdpune.gov.in/Links/annual%20summary%202016.pdf Accessed 1 Sept 2017

IMD (India Meteorological Department) (2019) Annual climate summary. Available via https://www.imdpune.gov.in/Links/annual_summary_2019.pdf. Accessed 1 Sept 2017

Immerzeel WW, Wanders N, Lutz AF, Shea JM, Bierkens MF (2015) Reconciling high-altitude precipitation in the upper Indus basin with glacier mass balances and runoff. Hydrol Earth Syst Sci 19(11):4673–4687

IPCC (2007) Climate change 2007: impacts, adaptation and vulnerability. In: Parry ML, Canziani OF, Palutik JP, van der Linden PJ, Hanson CE (eds) Contribution of working group II to the 4th assessment report of the intergovernmental panel on climate change. Cambridge UniversityPress, Cambridge

India-WRIS (2012) River basin atlas of India. RRSC-West, NRSC, ISRO, Jodhpur. Available via http://indiawris.gov.in/downloads/RiverBasinAtlas_Full.pdf

IPCC (2013) Climate change 2013: The physical science basis. Contribution of Working Group I to the Fifth Assessment Report of the Intergovernmental Panel on Climate Change (IPCC). Cambridge University Press, Cambridge

IPCC (2021a) Climate change 2021: the physical science basis. Contribution of Working Group 1 to Sixth Assessment Report of the Intergovernmental Panel on Climate Change. Available at: https://www.ipcc.ch/report/ar6/wg1/

IPCC (2021b) Climate change widespread, rapid, and intensifying – IPCC. IPCC press release AR6, (August 2021), pp 1–6

Ireson A, Makropoulos C, Maksimovic C (2006) Water resources modelling under data scarcity: coupling MIKE BASIN and ASM groundwater model. Water Resour Manag 20(4):567–590

Jabbar A, Othman AA, Merkel B, Hasan SE (2020) Change detection of glaciers and snow cover and temperature using remote sensing and GIS: a case study of the Upper Indus Basin, Pakistan. Remote Sens Appl: Soc Env 18:100308

Jasrotia AS, Baru D, Kour R, Ahmad S, Kour K (2021) Hydrological modeling to simulate stream flow under changing climate conditions in Jhelum catchment, western Himalaya. J Hydrol 593:125887

Jhajharia D, Singh VP (2011) Trends in temperature, diurnal temperature range and sunshine duration in Northeast India. Int J Climatol 31(9):1353–1367

Julien PY, Saghafian B, Ogden FL (1995) Raster-based hydrologic modelling of spatially-varied surface runoff. Water Resour Bull 31(3):523–536

Kääb A, Treichler D, Nuth C, Berthier E (2015) Contending estimates of 2003–2008 glacier mass balance over the Pamir–Karakoram–Himalaya [Brief Communication]. Cryosphere 9(2):557–564. https://doi.org/10.5194/tc-9-557-2015

Kapnick SB, Delworth TL, Ashfaq M, Malyshev S, Milly PCD (2014) Snowfall less sensitive to warming in Karakoram than in Himalayas due to a unique seasonal cycle. Nat Geosci 7:834–840

Kathuria HV, Thakur KK (2004) Catchment area treatment plan of Larji hydroelectric project. Mandi Forest Division, Mandi. https://agisac.gov.in/hpcampa/CD%20INDI%20CAT%20Plan/Lariji.pdf

Kendall M (1975) Rank correlation methods. Charles Griffin, London

Khan AJ, Koch M, Tahir AA (2020) Impacts of climate change on the water availability, seasonality and extremes in the Upper Indus Basin (UIB). Sustainability 12(4):1283

Khattak MS, Babel MS, Sharif M (2011) Hydro-meteorological trends in the upper Indus River basin in Pakistan. Clim Res 46(2):103–119

Kiani RS, Ali S, Ashfaq M, Khan F, Muhammad S, Reboita MS, Farooqi A (2021) Hydrological projections over the Upper Indus Basin at 1.5 °C and 2.0 °C temperature increase. Sci Total Environ 788:147759

Klein Tank AM, Peterson TC, Quadir DA, Dorji S, Zou X, Tang H, Santhosh K, Joshi UR, Jaswal AK, Kolli RK, Sikder AB (2006) Changes in daily temperature and precipitation extremes in central and south Asia. J Geophys Res Atmos 111(D16)

Kraaijenbrink PD, Bierkens MF, Lutz AF, Immerzeel WW (2017) Impact of a global temperature rise of 1.5 degrees Celsius on Asia's glaciers. Nature 549(7671):257–260

Krakauer NY, Lakhankar T, Dars GH (2019) Precipitation trends over the Indus basin. Climate 7(10):116

Krause P (2002) Quantifying the impact of land use changes on the water balance of large catchments using the J2000 model. Phys Chem Earth A/B/C 27(9–10):663–673

Krishnan R, Shrestha AB, Ren G, Rajbhandari R, Saeed S, Sanjay J, Syed M, Vellore R, Xu Y, You Q, Ren Y (2019) Unravelling climate change in the Hindu Kush Himalaya: rapid warming in the mountains and increasing extremes. In: The Hindu Kush Himalaya assessment. Springer, Cham, pp 57–97. https://doi.org/10.1007/978-3-319-92288-1_3

Krusell P, Smith Jr AA. (2022) Climate change around the world (No. w30338). National Bureau of Economic Research. https://doi.org/10.3386/w30338. https://www.nber.org/system/files/working_papers/w30338/w30338.pdf

Krysanova V, Müller-Wohlfeil DI, Becker A (1998) Development and test of a spatially distributed hydrological/water quality model for mesoscale watersheds. Ecol Model 106(2–3):261–289

Kulkarni B (2013) Hydrological response to climate change in the Krishna Basin. In: 2012 International SWAT conference proceedings. Indian Institute of Technology, Delhi, pp 80–87. Available via http://swat.tamu.edu/media/69009/swat-proceedings-2012-india.pdf. Accessed 20 May 2014

Kulkarni AV, Rathore BP, Mahajan S, Mathur P (2005) Alarming retreat of Parbati glacier, Beas basin, Himachal Pradesh. Curr Sci 10:1844–1850

Kumar R, Areendran G, Rao P (2009) Witnessing change: glaciers in the Indian Himalayas. WWF India and BIT. Available via www.wwf.se/source.php?id=1271034. Accessed 20 May 2014

Kumar V, Jain SK, Singh Y (2010) Analysis of long-term rainfall trends in India. Hydrol Sci J 55(4):484–496

Kumar P, Wiltshire A, Mathison C, Asharaf S, Ahrens B, Lucas-Picher P, Christensen JH, Gobiet A, Saeed F, Hagemann S, Jacob D (2013) Downscaled climate change projections with uncertainty assessment over India using a high resolution multi-model approach. Sci Total Environ 468:S18–S30

Latif Y, Yaoming M, Yaseen M (2018) Spatial analysis of precipitation time series over the Upper Indus Basin. Theor Appl Climatol 131(1):761–775

Latif Y, Ma Y, Ma W (2021) Climatic trends variability and concerning flow regime of Upper Indus Basin, Jehlum, and Kabul river basins Pakistan. Theor Appl Climatol 144(1):447–468

Li L, Shen M, Hou Y, Xu CY, Lutz AF, Chen J, Jain SK, Li J, Chen H (2019a) Twenty-first-century glacio-hydrological changes in the Himalayan headwater Beas River basin. Hydrol Earth Syst Sci Discuss 23(3):1483–1503. https://doi.org/10.5194/hess-23-1483-2019

Li Y, Chang J, Luo L, Wang Y, Guo A, Ma F, Fan J (2019b) Spatiotemporal impacts of land use land cover changes on hydrology from the mechanism perspective using SWAT model with time-varying parameters. Hydrol Res 50(1):244–261. https://doi.org/10.2166/nh.2018.006

Liaqat MU, Grossi G, Ranzi R (2022) Characterization of interannual and seasonal variability of hydro-climatic trends in the Upper Indus Basin. Theor Appl Climatol 147(3):1163–1184

Liu X, Cheng Z, Yan L, Yin ZY (2009) Elevation dependency of recent and future minimum surface air temperature trends in the Tibetan Plateau and its surroundings. Glob Planet Chang 68(3):164

Lu N, Qin J, Gao Y, Yang K, Trenberth KE, Gehne M, Zhu Y (2015) Trends and variability in atmospheric precipitable water over the Tibetan Plateau for 2000–2010. Int J Climatol 35(7):1394–1404

Liu J, Wang Z, Gong T (2012) Comparative analysis of hydroclimatic changes in glacier-fed rivers in the Tibet- and Bhutan-Himalayas. Quat Int 282:104–112.

Madhura RK, Krishnan R, Revadekar JV, Mujumdar M, Goswami BN (2015) Changes in western disturbances over the Western Himalayas in a warming environment. Clim Dyn 44:1157–1168. https://doi.org/10.1007/s00382-014-2166-9

Mahala A (2020) The significance of morphometric analysis to understand the hydrological and morphological characteristics in two different morpho-climatic settings. Appl Water Sci 10(1):1–6

Mal S, Singh RB, Huggel C, Grover A (2018) Introducing linkages between climate change, extreme events, and disaster risk reduction. In: Mal S, Singh R, Huggel C (eds) Climate change, extreme events and disaster risk reduction, Sustainable development goals series. Springer, Cham. https://doi.org/10.1007/978-3-319-56469-2_1

Mal S, Rani S, Maharana P (2020) Estimation of spatio-temporal variability in land surface temperature over the Ganga River Basin using MODIS data. Geocarto Int:1–23

Mal S, Arora M, Banerjee A, Singh RB, Scott CA, Allen SK, Karki R (2022) Spatial variations and long-term trends (1901–2013) of rainfall across Uttarakhand Himalaya, India. In: Schickhoff U, Singh R, Mal S (eds) Mountain landscapes in transition, Sustainable development goals series. Springer, Cham, pp 163–183. https://doi.org/10.1007/978-3-030-70238-0_3

Mann H (1945) Non-parametric test against trend. Econometrica 13:245–259

Martinec J (1975) Subsurface flow from snowmelt traced by tritium. Water Resour Res 11(3):496–498

Maxwell JC (1955) The bifurcation ratio in Horton's law of stream numbers. Trans Am Geophys Union 36:520.

Mayewski PA, Perry LB, Matthews T, Birkel SD (2020) Climate change in the Hindu Kush Himalayas: basis and gaps. One Earth 3(5):551–555. https://doi.org/10.1016/j.oneear.2020.10.007

Miller VC (1953) A quantitative geomorphologic study of drainage basin characteristics in the Clinch Mountain Area, Virginia and Tennessee. Technical report 3. Department of Geology Columbia University

Mishra V, Lilhare R (2016) Hydrologic sensitivity of Indian sub-continental river basins to climate change. Glob Planet Chang 139:78–96. https://doi.org/10.1016/j.gloplacha.2016.01.003

Mishra PK, Rai A, Rai SC (2020) Land use and land cover change detection using geospatial techniques in the Sikkim Himalaya, India. Egypt J Remote Sens Space Sci 23(2):133–143

Moglen GE, Eltahir EA, Bras RL (1998) On the sensitivity of drainage density to climate change. Water Resour Res 34(4):855–862. https://doi.org/10.1029/97wr02709

Mondal PP, Zhang Y (2018) Research progress on changes in land use and land cover in the western Himalayas (India) and effects on ecosystem services. Sustainability 10(12):4504. https://doi.org/10.3390/su10124504

Morán-Tejeda E, Zabalza J, Rahman K, Gago-Silva A, López-Moreno JI, Vicente-Serrano S, Lehmann A, Tague CL, Beniston M (2015) Hydrological impacts of climate and land-use changes in a mountain watershed: uncertainty estimation based on model comparison. Ecohydrology 8(8):1396–1416. https://doi.org/10.1002/eco.1590

Mukhopadhyay B, Khan A (2014) A quantitative assessment of the genetic sources of the hydrologic flow regimes in Upper Indus Basin and its significance in a changing climate. J Hydrol 509:549–572

Müller J, Dame J, Nüsser M (2020) Urban Mountain waterscapes: the transformation of hydrosocial relations in the trans-Himalayan town Leh, Ladakh, India. Water 12(6):1698

Nag SK, Chakraborty S (2003) Influence of rock types and structures in the development of drainage network in hard rock area. J Indian Soc Remote Sens 31(1):25–35. https://doi.org/10.1007/bf03030749

NATCOM- India's National Communications (2012) India second national communication to the United Nations framework convention on climate change. Available via http://www.undp.org/content/dam/india/docs/united_nations_framework_convention_on_climate_change.pdf

Nazeer A, Maskey S, Skaugen T, McClain ME (2022) Changes in the hydro-climatic regime of the Hunza Basin in the Upper Indus under CMIP6 climate change projections. Sci Rep 12(1):1–6

Neitsch SL, Arnold JG, Kiniry JR, Srinivasan R, Williams JR (2002) Soil and water assessment tool, user's manual, version 2000. Available via http://swat.tamu.edu/media/1294/swatuserman.pdf. Accessed 20 May 2014

Neitsch S, Arnold J, Kiniry J, Williams J (2011) Soil and water assessment tool- theoretical documentation version 2009. Texas Water resources Institute, Texas. Available via http://swat.tamu.edu/media/99192/swat2009-theory.pdf. Accessed 20 May 2014

Nepal S (2016) Impacts of climate change on the hydrological regime of the Koshi river basin in the Himalayan region. J Hydro-Environ Res 10:76–89. https://doi.org/10.1016/j.jher.2015.12.001

Nepal S, Shrestha AB (2015) Impact of climate change on the hydrological regime of the Indus, Ganges and Brahmaputra River basins: a review of the literature. Int J Water Resour Dev 31(2):201–218. https://doi.org/10.1080/07900627.2015.1030494

Neupane RP, White JD, Alexander SE (2015) Projected hydrologic changes in monsoon-dominated Himalaya Mountain basins with changing climate and deforestation. J Hydrol 525:216–230. https://doi.org/10.1016/j.jhydrol.2015.03.048

Nie Y, Liu Q, Wang J, Zhang Y, Sheng Y, Liu S (2018) An inventory of historical glacial lake outburst floods in the Himalayas based on remote sensing observations and geomorphological analysis. Geomorphology 308:91–106

Nie Y, Pritchard HD, Liu Q, Hennig T, Wang W, Wang X, Liu S, Nepal S, Samyn D, Hewitt K, Chen X (2021) Glacial change and hydrological implications in the Himalaya and Karakoram. Nat Rev Earth Environ 2(2):91–106

Nijssen B, Haddeland I, Lettenmaier DP (1997) Point evaluation of a surface hydrology model for BOREAS. J Geophys Res Atmos 102(D24):29367–29378

Nilawar AP, Waikar ML (2018) Use of SWAT to determine the effects of climate and land use changes on streamflow and sediment concentration in the Purna River basin, India. Environ Earth Sci 77(23):1–3

NRSC and ISRO (2011) Manual on "Preparation of geo spatial layers using high resolution (Cartosat-1Pan+LISS-IV Mx) orthorectified satellite imagery". Space Based Information Support for Decentralized Planning (SIS-DP), Remote Sensing and GIS Applications Area National Remote Sensing Centre, Indian Space Research Organisation (ISRO), Department of Space, Government of India, Hyderabad

Nüsser M, Schmidt S (2021) Glacier changes on the Nanga Parbat 1856–2020: a multi-source retrospective analysis. Sci Total Environ 785:147321

Nüsser M, Dame J, Parveen S, Kraus B, Baghel R, Schmidt S (2019) Cryosphere-fed irrigation networks in the northwestern Himalaya: precarious livelihoods and adaptation strategies under the impact of climate change. Mt Res Dev 39(2):R1

Osei MA, Amekudzi LK, Wemegah DD, Preko K, Gyawu ES, Obiri-Danso K (2019) The impact of climate and land-use changes on the hydrological processes of Owabi catchment from SWAT analysis. Environ Earth Sci 25:100620

Ougahi JH, Cutler ME, Cook SJ (2022) Modelling climate change impact on water resources of the Upper Indus Basin. J Water Clim Chang 13(2):482–504

Pal I, Al-Tabbaa A (2011) Assessing seasonal precipitation trends in India using parametric and non-parametric statistical techniques. Theor Appl Climatol 103(1):1–1

Pant GB, Borgaonkar HP, Rupa Kumar K (1998) Climatic signals from tree-rings: a dendro climatic investigation of Himalayan spruce (Picea smithiana). Himal Geol 19(2):65–73

Paul S, Ghosh S, Oglesby R, Pathak A, Chandrasekharan A, Ramsankaran RA (2016) Weakening of Indian summer monsoon rainfall due to changes in land use land cover. Sci Rep 6(1):1. https://doi.org/10.1038/srep32177

Pepin N, Bradley RS, Diaz HF, Baraer M, Caceres EB, Forsythe N, Fowler H, Greenwood G, Hashmi MZ, Liu XD, Miller JR (2015) Elevation-dependent warming in mountain regions of the world. Nat Clim Chang 5:424–430

Pervez MS, Henebry GM (2015) Assessing the impacts of climate and land use and land cover change on the freshwater availability in the Brahmaputra River basin. J Hydrol Reg Stud 3:285–311. https://doi.org/10.1016/j.ejrh.2014.09.003

Pokhrel Y, Burbano M, Roush J, Kang H, Sridhar V, Hyndman DW (2018) A review of the integrated effects of changing climate, land use, and dams on Mekong river hydrology. Water 10(3):266. https://doi.org/10.3390/w10030266

Pörtner HO, Roberts DC, Adams H, Adler C, Aldunce P, Ali E, Begum RA, Betts R, Kerr RB, Biesbroek R, Birkmann J (2022) Climate change 2022: impacts, adaptation and vulnerability. IPCC Sixth Assessment Report.

Pritchard HD (2019) Asia's shrinking glaciers protect large populations from drought stress. Nature 569(7758):649–654

Rai RK, Upadhyay A, Ojha CSP (2010) Temporal variability of climatic parameters of yamuna river basin: spatial analysis of persistence, trend and periodicity. Open J Hydrol 4:184–210

Rajbhandari R, Shrestha AB, Kulkarni A, Patwardhan SK, Bajracharya SR (2015) Projected changes in climate over the Indus river basin using a high resolution regional climate model (PRECIS). Clim Dyn 44(1):339–337

Ramzan S, Ahmed P, Mahmood R, Dimri AP (2019) Assessment of present and future climate change over Kashmir Himalayas, India. Theor Appl Climatol 137(3):3183–3195

Rani S (2014) Assessment of the influence of climate variability on the snow cover area of the Upper Beas River Basin. Unpublished M.Phil. Dissertation, CSRD, JNU

Rani S, Mal S (2022) Trends in land surface temperature and its drivers over the High Mountain Asia. Egypt J Remote Sens Space Sci 25(3):717–729

Rani S, Sreekesh S (2018) Variability of temperature and rainfall in the Upper Beas Basin, Western Himalaya. In: Climate change, extreme events and disaster risk reduction. Springer, Cham, pp 101–120. https://doi.org/10.1007/978-3-319-56469-2_7

Rani S, Sreekesh S (2021) Flow regime changes under future climate and land cover scenarios in the Upper Beas basin of Himalaya using SWAT model. Int J Environ Stud 78(3):382–397. https://doi.org/10.1080/00207233.2020.1811574

Rani S, Sreekesh S (2022) Assessment and prediction of land use/land cover changes of Beas basin using a modeling approach. In: Mountain landscapes in transition. Springer, Cham, pp 471–487

Rani S, Sreekesh S, Krishnan P (2019) Effect of climate change on potential evapotranspiration in the upper Beas basin of the western Indian Himalaya. Int Arch Photogramm Remote Sens Spat Inf Sci ISPRS Arch 42(3/W6):51–57. https://doi.org/10.5194/isprs-archives-XLII-3-W6-51-2019

Rani S, Kumar R, Maharana P (2022) Climate change, its impacts, and sustainability issues in the Indian Himalaya: an introduction. In: Climate change. Springer, Cham, pp 1–27

Ray DK, West PC, Clark M, Gerber JS, Prishchepov AV, Chatterjee S (2019) Climate change has likely already affected global food production. PLoS One 14(5):e0217148

Ren YY, Ren GY, Sun XB, Shrestha AB, You QL, Zhan YJ, Rajbhandari R, Zhang PF, Wen KM (2017) Observed changes in surface air temperature and precipitation in the Hindu Kush Himalayan region over the last 100-plus years. Adv Clim Chang Res 8(3):148–156

Rode M, Lindenschmidt K-E (2001) Distributed sediment and phosphorus transport modeling on a medium sized catchment in central germany. Phys Chem Earth B Hydrol Oceans Atmos 26(7–8):635–640

Romshoo SA, Altaf S, Rashid I, Dar RA (2018) Climatic, geomorphic and anthropogenic drivers of the 2014 extreme flooding in the Jhelum basin of Kashmir, India. Geomatics Natural Hazards Risk 9(1):224–248. https://doi.org/10.1080/19475705.2017.1417332

Romshoo SA, Murtaza KO, Abdullah T (2022) Towards understanding various influences on mass balance of the Hoksar Glacier in the Upper Indus Basin using observations. Sci Rep 12(1):1–4

Sahu R, Gupta RD (2020) Glacier mapping and change analysis in Chandra basin, Western Himalaya, India during 1971–2016. Int J Remote Sens 41(18):6914–6945

Sarkar D, Mondal P, Sutradhar S, Sarkar P (2020) Morphometric analysis using SRTM-DEM and GIS of Nagar river basin, Indo-Bangladesh barind tract. J Indian Soc Remote Sens 48(4):597–614. https://doi.org/10.1007/s12524-020-01106-7

Sarthi PP, Dash SK, Mamgain A (2012) Possible changes in the characteristics of Indian Summer Monsoon under warmer climate. Glob Planet Chang 92:17–29

Saxena R, Mathur P (2019) Recent trends in rainfall and temperature over North West India during 1871–2016. Theor Appl Climatol 135(3):1323–1338

Saxton KE, Rawls WJ (2006) Soil water characteristic estimates by texture and organic matter for hydrologic solutions. Soil Sci Soc Am J 70(5):1569–1578

Saxton KE, Rawls W, Romberger JS, Papendick RI (1986) Estimating generalized soil-water characteristics from texture. Soil Sci Soc Am J 50(4):1031–1036

Schiavina M, Freire S, MacManus K (2019) GHS population grid multitemporal (1975-1990-2000-2015), R2019A. European Commission, Joint Research Centre (JRC) [Dataset]. https://doi.org/10.2905/0C6B9751-A71F-4062-830B-43C9F432370F. PID: http://data.europa.eu/89h/0c6b9751-a71f-4062-830b-43c9f432370f

Schickhoff U, Singh RB, Mal S (2016) Climate change and dynamics of glaciers and vegetation in the Himalaya: an overview. In: Singh R, Schickhoff U, Mal S (eds) Climate change, glacier response, and vegetation dynamics in the Himalaya. Springer, Cham. https://doi.org/10.1007/978-3-319-28977-9_1

Schimming CG, Mette R, Reiche EW, Schrautzer J, Wetzel H (1995) Nitrogen fluxes in a typical agroecosystem in Schleswig-Holstein-measurements, budgets, model validation. Zeitschrift fur Pflanzenernahrung und Bodenkunde 158:313

Schmidt S, Nüsser M (2012) Changes of high altitude glaciers from 1969 to 2010 in the Trans-Himalayan Kang Yatze Massif, Ladakh, northwest India. AAAR 44(1):107–121

Schmidt S, Nüsser M (2017) Changes of high altitude glaciers in the Trans-Himalaya of Ladakh over the past five decades (1969–2016). Geosciences 7(2):27

Schmidt S, Nüsser M, Baghel R, Dame J (2020) Cryosphere hazards in Ladakh: the 2014 Gya glacial lake outburst flood and its implications for risk assessment. Nat Hazards 104(3):2071–2095

Schumm SA (1956) Evolution of drainage systems and slopes in badlands at Perth Amboy, New Jersey. Geol Soc Am 67(5):597–646

Sen P (1968) Estimates of the regression coefficient based on Kendall's tau. J Am Stat Assoc 63(324):1379–1389

Sen SM, Kansal A (2019) Achieving water security in rural Indian Himalayas: a participatory account of challenges and potential solutions. J Environ Manag 245:398–408. https://doi.org/10.1016/j.jenvman.2019.05.132

Shah MI, Khan A, Akbar TA, Hassan QK, Khan AJ, Dewan A (2020) Predicting hydrologic responses to climate changes in highly glacierized and mountainous region Upper Indus Basin. R Soc Open Sci 7(8):191957

Shi P, Chen C, Srinivasan R, Zhang X, Cai T, Fang X, Qu S, Chen X, Li Q (2011) Evaluating the SWAT model for hydrological modeling in the Xixian watershed and a comparison with the XAJ model. Water Resour Manag 25(10):2595–2612

Shrestha UB, Gautam S, Bawa KS (2012) Widespread climate change in the Himalayas and associated changes in local ecosystems. PLoS ONE 7(5):e36741. https://doi.org/10.1371/journal.pone.0036741

Shekar M, Chand S, Kumar S, Srinivasan K, Ganju A (2010) Climate-change studies in the western Himalaya. Ann Glaciol 51(54):105–112

Shrestha AB, Agrawal NK, Alfthan B, Bajracharya SR, Maréchal J, van Oort B (eds) (2015) The Himalayan climate and water atlas: impact of climate change on water resources in five of Asia's major river basins. ICIMOD, GRID-Arendal and CICERO

Shukla A, Garg S, Mehta M, Kumar V, Shukla UK (2020) Temporal inventory of glaciers in the Suru sub-basin, western Himalaya: impacts of regional climate variability. Earth Syst Sci Data 12(2):1245–1265

Singh RB, Mal S (2014) Trends and variability of monsoon and other rainfall seasons in Western Himalaya, India. Atmos Sci Lett 15(3):218–226. https://doi.org/10.1002/asl2.494

Singh P, Thakur JK, Singh UC (2013) Morphometric analysis of Morar River Basin, Madhya Pradesh, India, using remote sensing and GIS techniques. Environ Earth Sci 68(7):1967–1977

Singh R, Kumar A, Kumar R (2014) Ecosystem services in changing environment. In: Singh R, Heitala R (eds) Livelihood security in northwestern Himalaya: case studies from changing socio-economic environments in Himachal Pradesh, India. Springer, Tokyo, pp 139–156

Singh S, Kumar R, Bhardwaj A, Sam L, Shekhar M, Singh A, Kumar R, Gupta A (2016) Changing climate and glacio-hydrology in Indian Himalayan Region: a review. Wiley Interdiscip Rev Clim Chang 7(3):393–410

SLUSI (2013) Soil resource mapping District Kullu and Mandi, Himachal Pradesh. Soil and Land Use of Survey India

Smith KG (1950) Standards for grading texture of erosional topography. Am J Sci 248(9):655–668

Stern N (2007) Stern review on the economics of climate change. Cambridge University Press, Cambridge

Strahler AN (1952) Hypsometric analysis (area-altitude) of erosional topography. Geol Soc Am Bull 63:117–142

Strahler AN (1957) Quantitative analysis of watershed geomorphology. Trans Am Geophys Union 38(6):913–920. https://doi.org/10.1029/TR038i006p00913

Strahler AN (1964) Quantitative geomorphology of drainage basin and channel networks. In: Handbook of applied hydrology. McGraw Hill Book Company, New York

Talib A, Randhir TO (2017) Climate change and land use impacts on hydrologic processes of watershed systems. J Water Clim Chang 8(3):363–374. https://doi.org/10.2166/wcc.2017.064

Tewari VP, Verma RK, von Gadow K (2017) Climate change effects in the Western Himalayan ecosystems of India: evidence and strategies. For Ecosyst 4:13. https://doi.org/10.1186/s40663-017-0100-4

Thayyen RJ, Gergan JT (2009) Role of glaciers in watershed hydrology: "Himalayan catchment" perspective. Cryosphere Discuss 3(2):443–476

Theil H (1950) A rank-invariant method of linear and polynomial regression analysis. Koninkluke Nederlandse Akademie Van Wetenschappen 53:467–482

Ullah S, Tahir AA, Akbar TA, Hassan QK, Dewan A, Khan AJ, Khan M (2019) Remote sensing-based quantification of the relationships between land use land cover changes and surface temperature over the Lower Himalayan Region. Sustainability 11(19):5492

United Nations (2019) World population prospects 2019: Department of economic and social Affairs. World Population Prospects 2019. https://population.un.org/wpp/

Vatsal S, Bhardwaj A, Azam MF, Mandal A, Ramanathan A, Bahuguna I, Raju NJ, Tomar SS (2022) A comprehensive multidecadal glacier inventory dataset for the Chandra-Bhaga Basin, Western Himalaya, India. Earth Syst Sci Data Discuss:1–38

Vishwas P (2021) Quantitative evaluation of drainage attributes to infer hydrologic and morphological characteristics of upper Beas Basin, Himachal Pradesh: a GIS-based approach. Geol Ecol Landsc:1–6. https://doi.org/10.1080/24749508.2021.1952766

Wagner PD, Bhallamudi SM, Narasimhan B, Kantakumar LN, Sudheer KP, Kumar S, Schneider K, Fiener P (2016) Dynamic integration of land use changes in a hydrologic assessment of a rapidly developing Indian catchment. Sci Total Environ 539:153–164. https://doi.org/10.1016/j.scitotenv.2015.08.148

Wang-Erlandsson L, Fetzer I, Keys PW, Van Der Ent RJ, Savenije HH, Gordon LJ (2018) Remote land use impacts on river flows through atmospheric teleconnections. Hydrol Earth Syst Sci 22(8):4311–4328. https://doi.org/10.5194/hess-22-4311-2018

Westra S, Alexander LV, Zwiers FW (2013) Global increasing trends in annual maximum daily precipitation. J Clim 26(11):3904–3918

Wijngaard RR, Lutz AF, Nepal S, Khanal S, Pradhananga S, Shrestha AB, Immerzeel WW (2017) Future changes in hydro-climatic extremes in the Upper Indus, Ganges, and Brahmaputra River basins. PLoS One 12(12):e0190224

Williams JR (1995) Chapter 25. The EPIC model. In: Computer models of watershed hydrology. Water Resources Publications, Highlands Ranch, pp 909–1000

World Water Council (2000) World water vision for the 21st century report. Available via www.worldwatercouncil.org/fileadmin/wwc/Library/WWVision/TableOfContents.pdf. Accessed 20 May 2014

Worni R, Huggel C, Stoffel M (2013) Glacial lakes in the Indian Himalayas – from an area-wide glacial lake inventory to on-site and modeling based risk assessment of critical glacial lakes. Sci Total Environ 468:S71–S84

Wu K, Liu S, Jiang Z, Liu Q, Zhu Y, Yi Y, Xie F, Tahir AA, Saifullah M (2021) Quantification of glacier mass budgets in the Karakoram region of Upper Indus Basin during the early twenty-first century. J Hydrol 603:127095

Xenarios S, Gafurov A, Schmidt-Vogt D, Sehring J, Manandhar S, Hergarten C, Shigaeva J, Foggin M (2019) Climate change and adaptation of mountain societies in Central Asia: uncertainties, knowledge gaps, and data constraints. Reg Environ Chang 19(5):1339–1352. https://doi.org/10.1007/s10113-018-1384-9

Yadav RR, Park WK, Bhattacharya A (1999) Spring temperature variations in the western Himalayan region as reconstructed from tree-rings: AD 1390–1987. The Holocene 9(1):85–90

Yadav RK, Kumar KR, Rajeevan M (2010) Climate change scenarios for Northwest India winter season. Quat Int 213(1–2):12–19

Yan J, Haan CT (1991) Multiobjective parameter estimation for hydrologic models-weighting of errors. Trans ASABE 34(1):135–141

Yarnal B, Lakhtakia MN, Yu Z, White RA, Pollard D, Miller DA, Lapenta WM (2000) A linked meteorological and hydrological model system: the Susquehanna River Basin Experiment (SRBEX). Glob Planet Chang 25(1–2):149–161

You QL, Ren GY, Zhang YQ, Ren YY, Sun XB, Zhan YJ, Shrestha AB, Krishnan R (2017) An overview of studies of observed climate change in the Hindu Kush Himalayan (HKH) region. Adv Clim Chang Res 8(3):141–147. https://doi.org/10.1016/j.accre.2017.04.001

Zhan YJ, Ren GY, Shrestha AB, Rajbhandari R, Ren YY, Sanjay J (2017) Change in extreme precipitation events over the HinduKush Himalayan region during 1961–2012. Adv Clim Chang Res 8(3). https://doi.org/10.1016/j.accre.2017.08.002

Zhang L, Wu L, Gan B (2013) Modes and mechanisms of global water vapor variability over the twentieth century. J Clim 26(15):5578–5593

Zhang H, Wang B, Li Liu D, Zhang M, Leslie LM, Yu Q (2020) Using an improved SWAT model to simulate hydrological responses to land use change: a case study of a catchment in tropical Australia. J Hydrol 585:124822

Zhao H, Su B, Lei H, Zhang T, Xiao C (2022) A new projection for glacier mass and runoff changes over High Mountain Asia. Sci Bull. https://doi.org/10.1016/j.scib.2022.12.004

Zhou G, Wei X, Chen X, Zhou P, Liu X, Xiao Y, Sun G, Scott DF, Zhou S, Han L, Su Y (2015) Global pattern for the effect of climate and land cover on water yield. Nat Commun 6(1):1–9

Chapter 2
Description of the Beas River Basin

Abstract The Beas River is also referred to as Arjikiya and Vipasha in Vedic and Sanskrit literature. It is the main river of India and one of the main tributaries of the Indus River. It covers an area of around 5.63% of the Indus River basin. It starts in the Beas Kund, which is situated in the Kullu district of the Himachal Pradesh at a height of ~4085 meters above mean sea level (AMSL), near the Rohtang Pass. The river has a total length of 460 km with a catchment area of 20303 km^2 (from the source to the destination). In terms of physiography, climate, soil, vegetation, hydrology, economics, and culture, the Beas River is a dynamic river system. Millions of people upstream and downstream rely on it for a living. The study of hydrological behavior is vital for the population. Compared to the high-altitude regions of the basin, the homogeneity is great in the lower basin. Along with providing a means of subsistence for a sizable population, it also faces problems with landslides, avalanches, and floods that need careful analysis. To predict the scenario in the future, it is necessary to comprehend the basin's current conditions.

Keywords Beas River Basin · Climate · Physical traits · People · Culture · Economy · Energy · Hazards

2.1 Physiography

The Beas River basin has complex physiography with diversity. The elevation in the basin varies from 126 m to more than 6000 m AMSL along the northeast edge of the Pārbati sub-catchment (Fig. 2.1). About 36% of the basin area is below 500 m of elevation. Additionally, 17% of the region has an elevation of more than 3000 m, while 2% of the basin has an elevation of more than 5500 m. High peaks may be seen both to the east and the north of the river valley. The highest peak of the basin is Parvati Parvat which is about 6600 m, located in the Kullu district (Parvati River basin) of Himachal Pradesh (Fig. 2.2). Other highest peaks of the basin are Dibibokri Pyramid, Indrasan, Umashila, Deo Tibba, Solang, Maiwa Kandinu, Hanuman

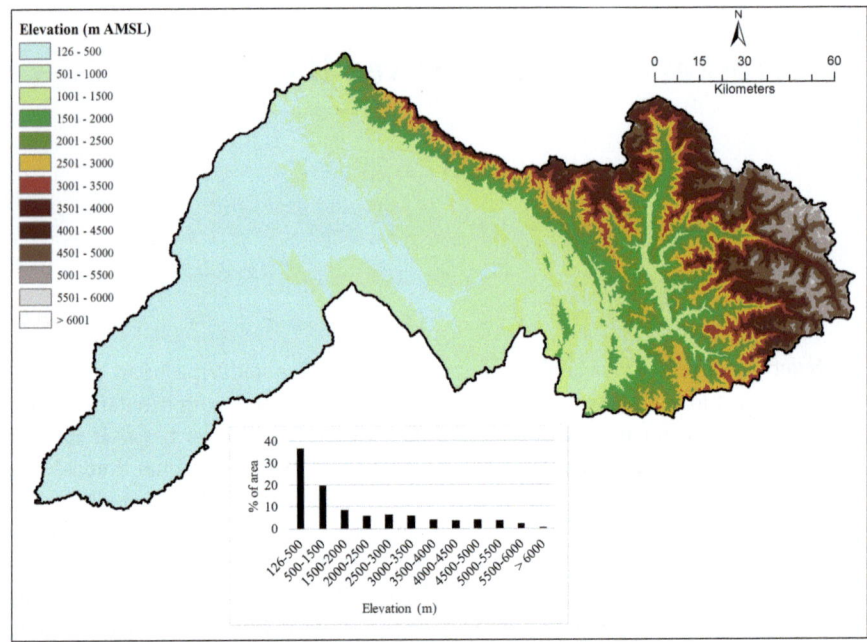

Fig. 2.1   Distribution of elevation in the Beas River basin

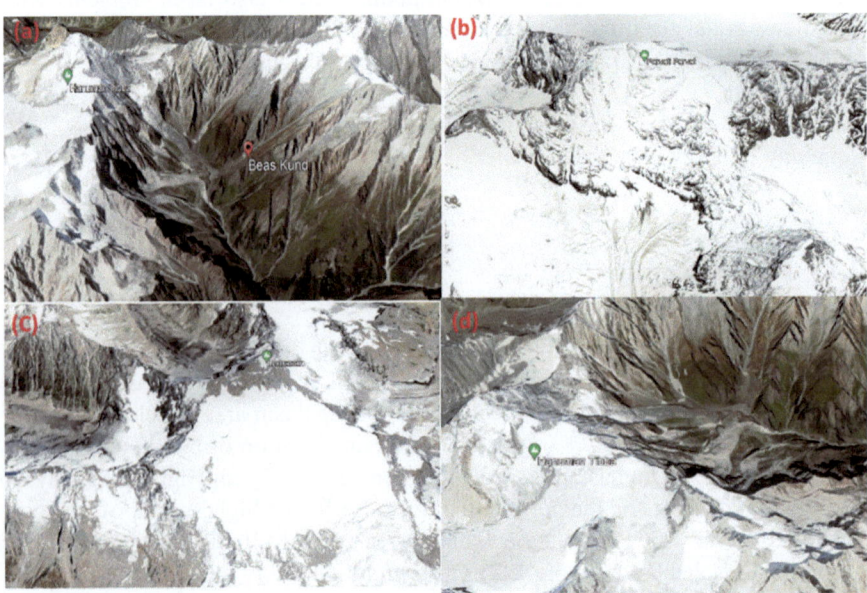

Fig. 2.2   Origin and main peaks: (**a**) Beas Kund (32° 21′59.49″N, 77° 5′8.08″E), (**b**) Parvati Parvat (32° 5′21.00″N, 77° 43′60.00″E), (**c**) Indrasan (32° 12′45.45″N, 77° 23′48.82″E), and (**d**) Hanuman Tibba (32° 20′6.31″N, 77° 3′2.58″E) of the Beas River basin, prepared from Google Earth (2019)

Tibba, Shitindhar, Srikhand Mahadev, Inderkilla, Patalsu, Shacha, etc. Geomorphic, tectonic, meteorological, and vegetation records may be found in the Pir Panjal range, which is also the source of the Beas River. Physiographically, the basin is very diverse and classified into (i) the Greater Himalayan Ranges (above 4500 m), (ii) the Middle Himalayan Ranges (1500–4500 m), (iii) the Shiwaliks (600–1500 m), (iv) the Kandi Region (300–400 m), and (v) the Alluvial Plains (<300 m) (Kumar and Rao 2021). The dramatic zonation in plant type, geomorphic processes, and related landforms is a result of the rising altitude. The principal valleys of the Beas River basin are Kullu (about 65 km), Parvati, Kangra, and Mandi valley (about 115 km) (Fig. 2.3). Some small valleys of the basin are Chauntra valley, Chauhar valley, and Balh valley.

The basin is largely comprised of precipitous slopes that vary from very gentle in the Punjab plains to steep slopes in the upper reaches of the basin in Himachal Pradesh (Fig. 2.4). More than half of the area of the basin has a slope of less than 15° while 25% of the area has a slope between 15° and 30°. About 4% of the basin's area has more than 45° slope.

The aspect of the basin shows almost uniform distribution (Fig. 2.5). About 13% of the area of the basin is found in the North, East, and South. More than 15% of the basin lies in the West, while 10–11% lies in the Southeast, Northeast, and Northwest. Elevation, slope, and aspect are decisive factors of weather, soil, vegetation, settlement, cultivation, and infrastructure development in Himachal Pradesh.

When compared to the glaciers and drainage from the Beas River to the south, which follow a southerly path, the Chandrabhaga runs east-west in the region north of the Pir Panjal. There are many significant passes in the basin, including Hamta Jot (4268 m), Haishin Jot (4942 m), Taintu Ka Jot (4996 m), Thanod Pass (4880 m), and Rohtang Pass (3998 m). Rivers, streams, nalas, springs, and glaciers all contribute to the region's water supply. Rawla (4500 m), Dhundha (4200 m), Ghoru, and Chandar (4800 m) are significant glaciers among many found in the area.

2.2 Geomorphology

The basin lies in the states of Himachal Pradesh and Punjab of India. Geological rocks vary from Protcrozoic to Quaternary representing classic geological sequences (Geological Survey of India 2012). Undifferentiated Proterozoic is mostly confined to the Lesser Himalayas. The Beas River includes features of structural origin, denudational origin, fluvial origin, glacial origin, and anthropogenic origin (https://bhuvan-app1.nrsc.gov.in/thematic/thematic/index.php#). The cross-section profile of the basin indicates the topographical variation with elevation difference from about 4000 to 2500 m (north–south) while from 5000 to 150 m (east–west) (Fig. 2.6).

The geomorphology of the basin in the upper reaches up to Pandoh Dam is comprised of structural origin-highly dissected hills and valleys, structural origin-moderately dissected hills and valleys, denudational origin-moderately dissected hills and valleys, denudational origin-Piedmont slope, denudational origin-mass

Fig. 2.3 Views of the Beas Valley around (**a–d**) Manali, (**e**) Naggar, and (**f**) Kullu cities

wasting products, fluvial origin-younger alluvial plain, fluvial origin-active flood plain, fluvial origin-Piedmont alluvial plain, glacial origin-glacial terrain, glacial origin-snow cover, and anthropogenic origin-anthropogenic terrain. The Beas River in the state of Punjab comprises features of structural origin, fluvial origin, aeolian origin, anthropogenic origin, and waterbodies (Fig. 2.7) (https://bhuvan-app1.nrsc. gov.in/thematic/thematic/index.php#).

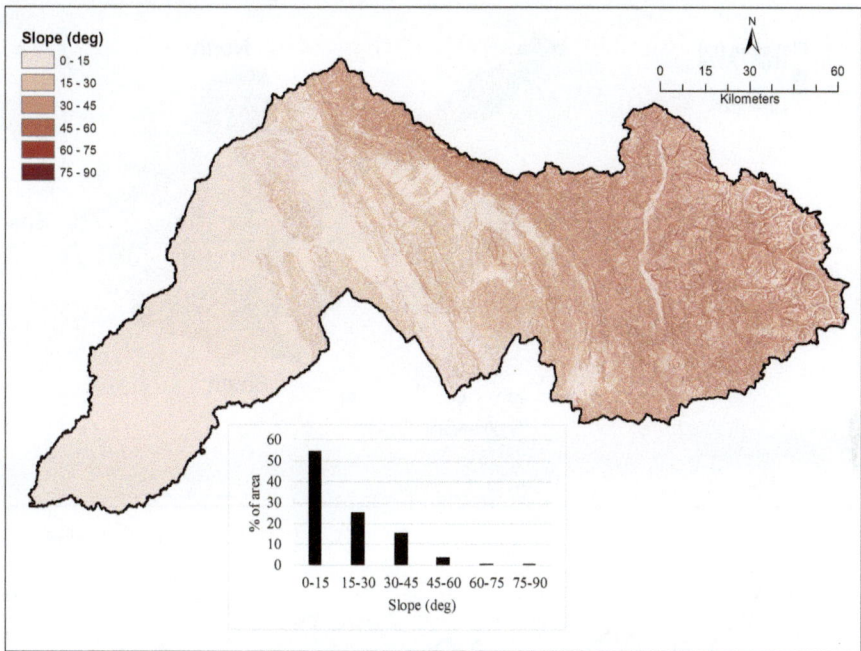

Fig. 2.4 Distribution of slope in the Beas River basin

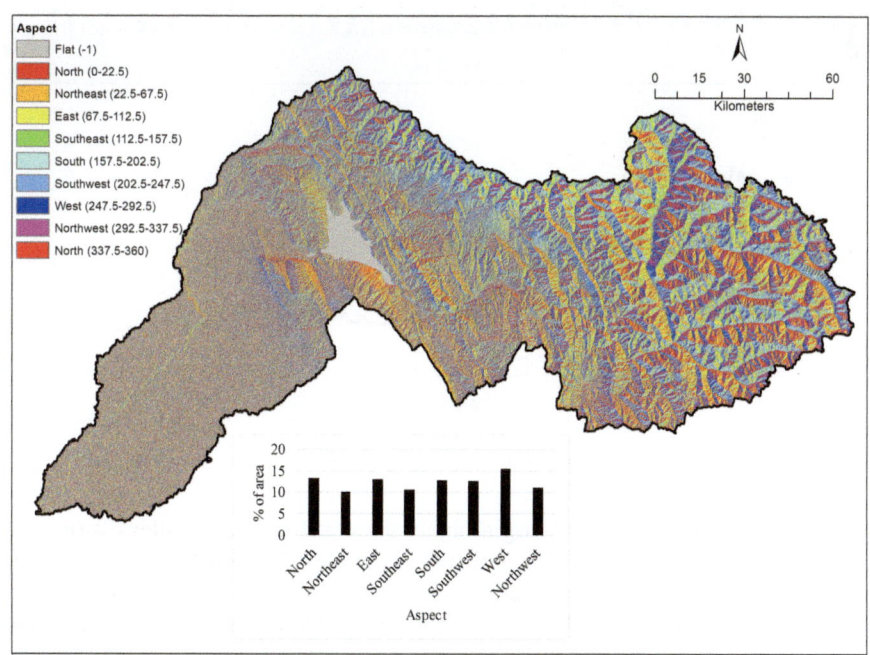

Fig. 2.5 Distribution of aspect in the Beas River basin

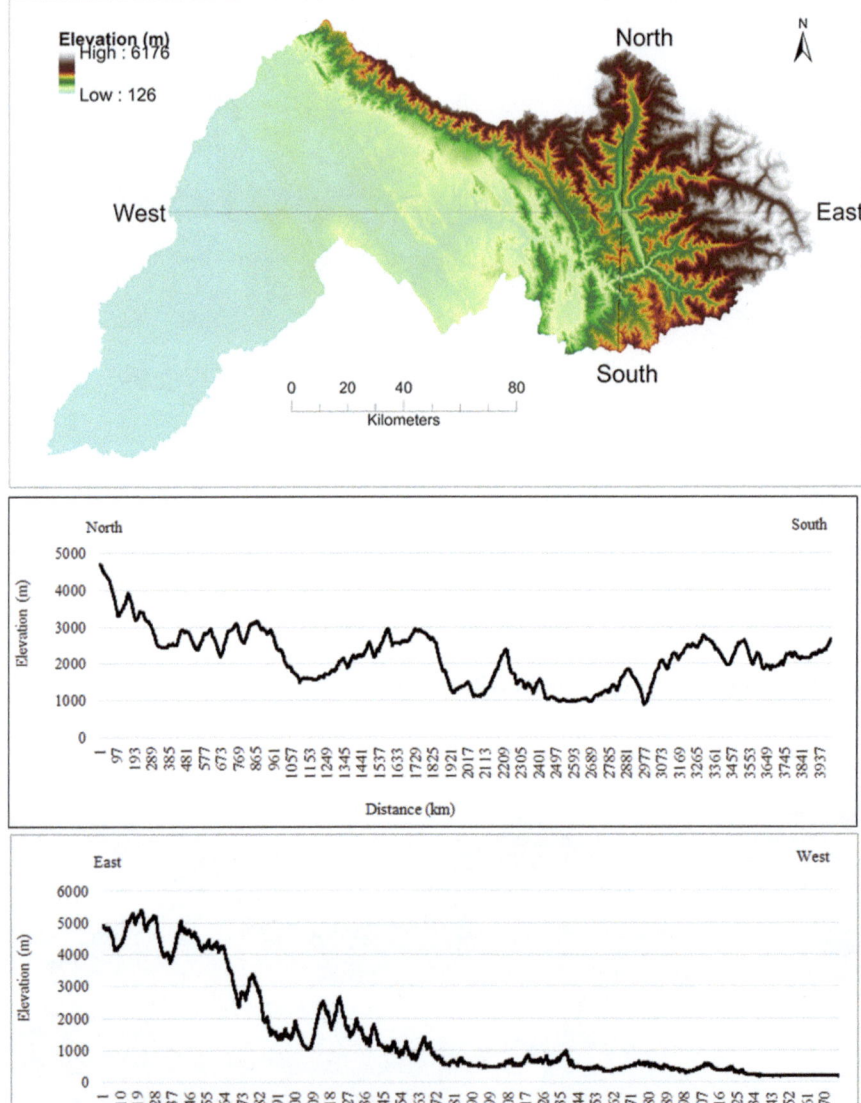

Fig. 2.6 Cross section profile of the Beas River basin

These consist of structural origin-highly dissected hills and valleys, structural origin-moderately dissected hills and valleys, structural origin-low dissected hills and valleys, fluvial origin-younger alluvial plain, fluvial origin-older flood plain, fluvial origin-active flood plain, fluvial origin-Piedmont alluvial plain, anthropogenic origin-anthropogenic terrain, and waterbodies. About 80% area of the basin in

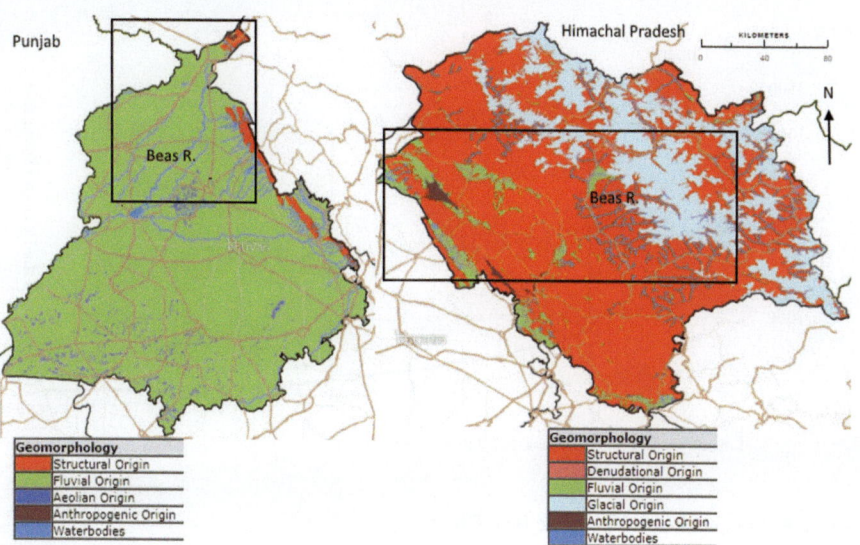

Fig. 2.7 Geomorphology of the Beas River basin, prepared from Bhuvan-Thematic Services (2005–2006)

the Punjab state comes under alluvial and flood plains. The Punjab state is forming part of the Indo-Gangetic basin and has two broad geomorphic entities, viz., the Siwalik foothills and the alluvial fill of the Indus drainage basin. The dominant physiographic divisions are "(i) Lahore-Sargodha Ridge in the west; (ii) Delhi-Jagadhari Ridge in the east; (iii) Delhi-Lahore Ridge in the south; and (iv) Siwalik ridges in the northeast."

2.3   Drainage

The Beas River originates at Beas Kund and gets its water from Manalsu Nalla, Allan Nalla, Phojal Nalla, Sarvai Nalla, Duhungan Nalla, and Mohal Nalla. The Parbati River near Bhuntar, the Tirthan and Sainj Rivers near Larji, the Sabari Nala near Kullu, and the Bakhli Khad close to Pandoh Dam are some of the important tributaries that joins the Beas River above Pandoh Dam. In Mandi district, it is joined by Hansa, Tirthan, Bakhli, Jiuni, Suketi, Panddi, Son, and Bather rivers. Among its tributaries, Parbati and Sainj Khad Rivers are glacier fed. The longitudinal profile of the Beas River is shown in Fig. 2.8. It indicates the differences in the height at the origin of the river and the Beas Satluj confluence. It is useful in understanding the young (left-hand side), middle, and old stages (right-hand side) of the river. The height of the river drops drastically in two places, namely, (i) Pandoh Dam and (ii) Mrithal.

The cross-section profile of the river indicates the behavior and landform features (V-shaped valley, U-shaped valley) in the young, middle, and old stages (Figs. 2.9 and 2.10).

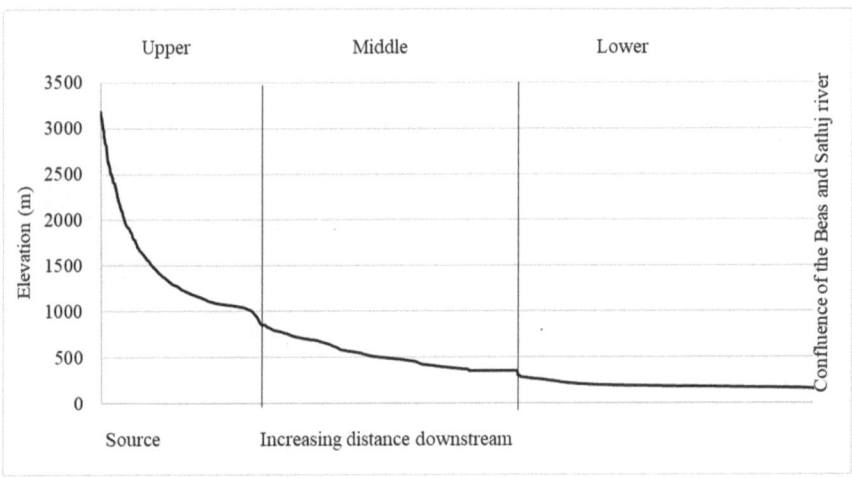

Fig. 2.8 Longitudinal profile of the Beas River

The basin has 21 different water bodies. The Beas River discharge fluctuates seasonally and annually. The monsoon season accounts for more than half of the Beas River's annual discharge, followed by the pre-monsoon, post-monsoon, and winter seasons (Kumar et al. 2007). Snowmelt accounts for around 25% of the Beas River's total annual flow. Snowfall and rain were estimated to have contributed to the total Beas River discharge at Pandoh Dam between 1990 and 2004 (Kumar et al. 2007). According to observations, rainfall contributes around 65% of the annual river discharge, with the remaining 35% coming from snow and glaciers combined. The ratio of rainfall and snowfall to total river discharge fluctuates significantly throughout time. The hydrology of the rivers is still influenced by seasonal factors, with winter (January–March) seeing the lowest flows, spring (April–May) seeing the highest flows due to snowmelt, and summer (June–September) seeing the highest discharges due to the combination of monsoon rainfall and runoff (Momblanch et al. 2019). Discharges eventually decline once more when rainfall drops throughout the post-monsoon season (October to December) (Momblanch et al. 2019). About 35% of the Beas River's annual flow at Pandoh Dam is a result of snowmelt and glacier runoff (Kumar et al. 2007). When the Pong Dam was constructed, seasonal variations in monthly releases decreased, and the lowest average monthly flow rose (120–300 m³/s) (Vercruysse and Grabowski 2021).

The Bhakra Beas Management Board (BBMB) replaced the Bhakra Management Board (BMB), which had been established in 1966 (located in Chandigarh union territory of India). It controls the distribution of water and electricity (administration, operation, and maintenance) to Punjab, Haryana, Rajasthan, Himachal Pradesh, New Delhi, and Chandigarh from the Bhakra Nangal and Beas Projects. It serves a further purpose in the development of new hydro projects both inside and outside the BBMB system.

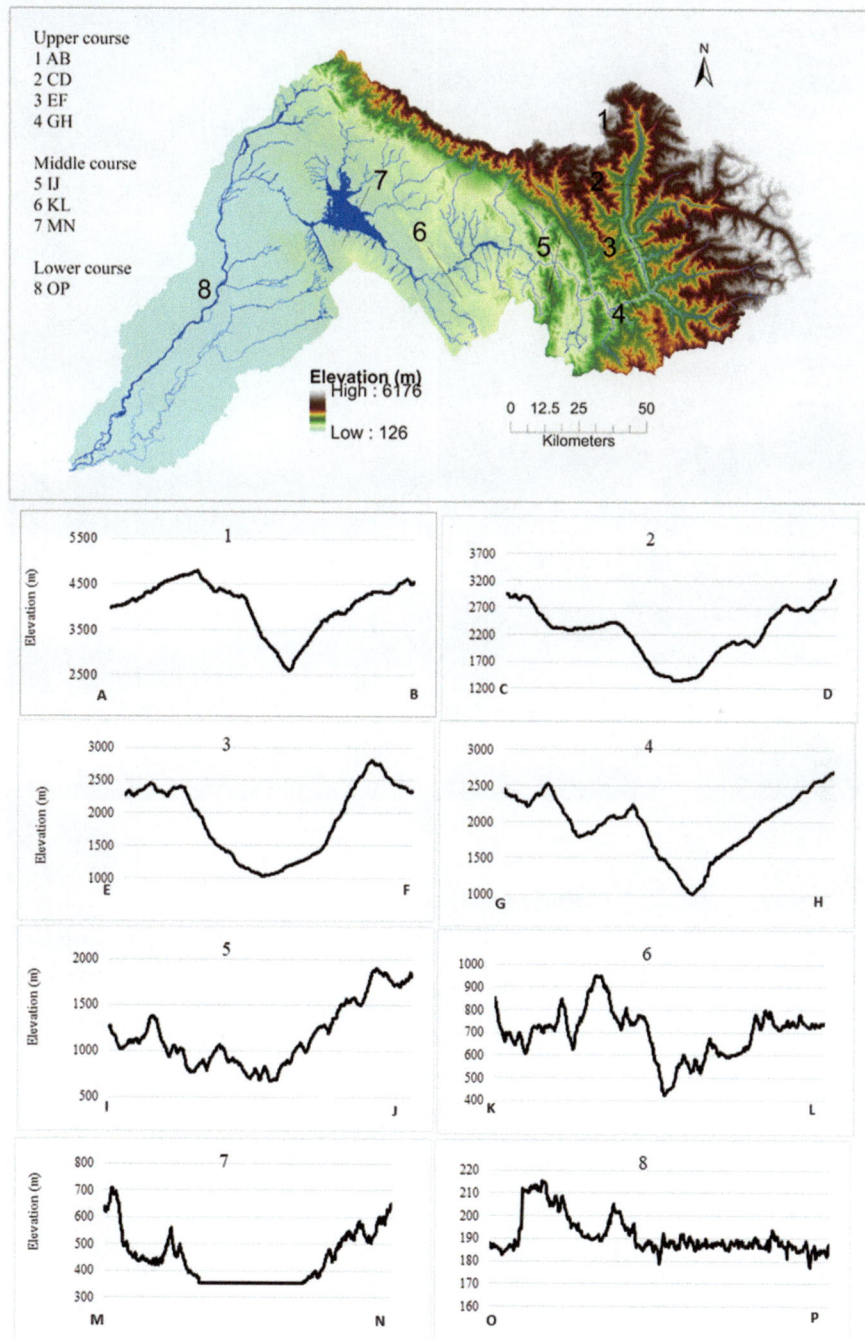

Fig. 2.9 Cross section profiles of the Beas River

Fig. 2.10 (1) Beas few kms ahead of Dhundi on way to Beas Kund, (2) Beas few kms ahead of Dhundi on way to Beas Kund, (3) Beas R. at Dhundi, (4) Beas R. near Dhundi bridge, (5) Beas R. at Manali, (6) Baragran, 15 miles near Manali, (7) Beas R. at Bhuntar, (8) Beas R. at Thalout, (9) Parbati R. at Pulga Dam, (10) Parbati River near Manikaran (fieldwork May–June 2017)

2.3.1 Morphometric Characteristics

Morphometric characteristics of a basin help to understand its hydrological condition. It includes stream order, length, basin area, etc. Stream order is based on Strahler's scheme of order in which each fingertip is allotted first order and two similar order streams converged to form a higher-order stream, and if two dissimilar order streams combine, the order carries the value of higher order in between two streams. This is the first and the basic step toward the identification of different morphometric parameters. It gives an idea of the position and association of one stream with another. The stream order of the basin depends on the form, extent, and relief traits of an area (Mahala 2020). Stream number, a high number of lower-order streams, denotes the mountainous region and young topography.

The Beas basin is a sixth-order drainage basin containing a total of 1010 streams of all orders, out of which 787 are first-, 172 are second-, 39 are third-, 9 are fourth-, 2 are fifth-, and 1 are sixth-order streams (Table 2.1 and Fig. 2.11). The sudden decrease in the third to fourth order stream is attributed to major morphological adjustments (Mahala 2020). The stream number goes on decreasing with decreasing height of the basin, which indicates the dominance of erosional landform, but there is an increment seen in the fifth-order stream, which is due to the difference in topography. The Beas basin from Beas Kund up to Aut is a sixth-order drainage basin with 1130 streams and streams from first to sixth order are 946, 140, 33, 7, 3, and 1, respectively (Vishwas 2021). The Parbati basin (sub-watershed of the Beas basin) is a sixth-order drainage basin with 3377 streams with streams, 2680, 562, 108, 21, 5, and 1, respectively, from first to sixth order (Khan et al. 2021). These two study areas are confined to the upper section of the Beas basin mainly upstream of Larji dam.

2.3.1.1 Total Stream Length (Lu)

It is useful in understanding the runoff characteristics of the basin. The total stream length decreases with stream order and follows the approximately linear pattern on graphs between stream order vs stream length, but a change in this general pattern indicates the fact there is a little discrepancy in the lithology due to high relief or moderately steep slopes which may arise due to varying lithology or probable uplift of the area (Khan et al. 2021). The total stream length of the Beas basin of first-,

Table 2.1 Estimated values of morphometric linear parameters of the Beas basin

Stream order	I	II	III	IV	V	VI	Total
Stream number	787	172	39	9	2	1	1010
Stream length (km)	2576.4	1366.3	662.3	242	203	52.3	5102.3
Mean stream length (km)	3.27	7.94	16.98	26.89	101.48	52.27	34.81
Stream length ratio		2.43	2.14	1.58	3.77	0.52	2.09
Bifurcation ratio	4.6	4.4	4.3	4.5	2		4

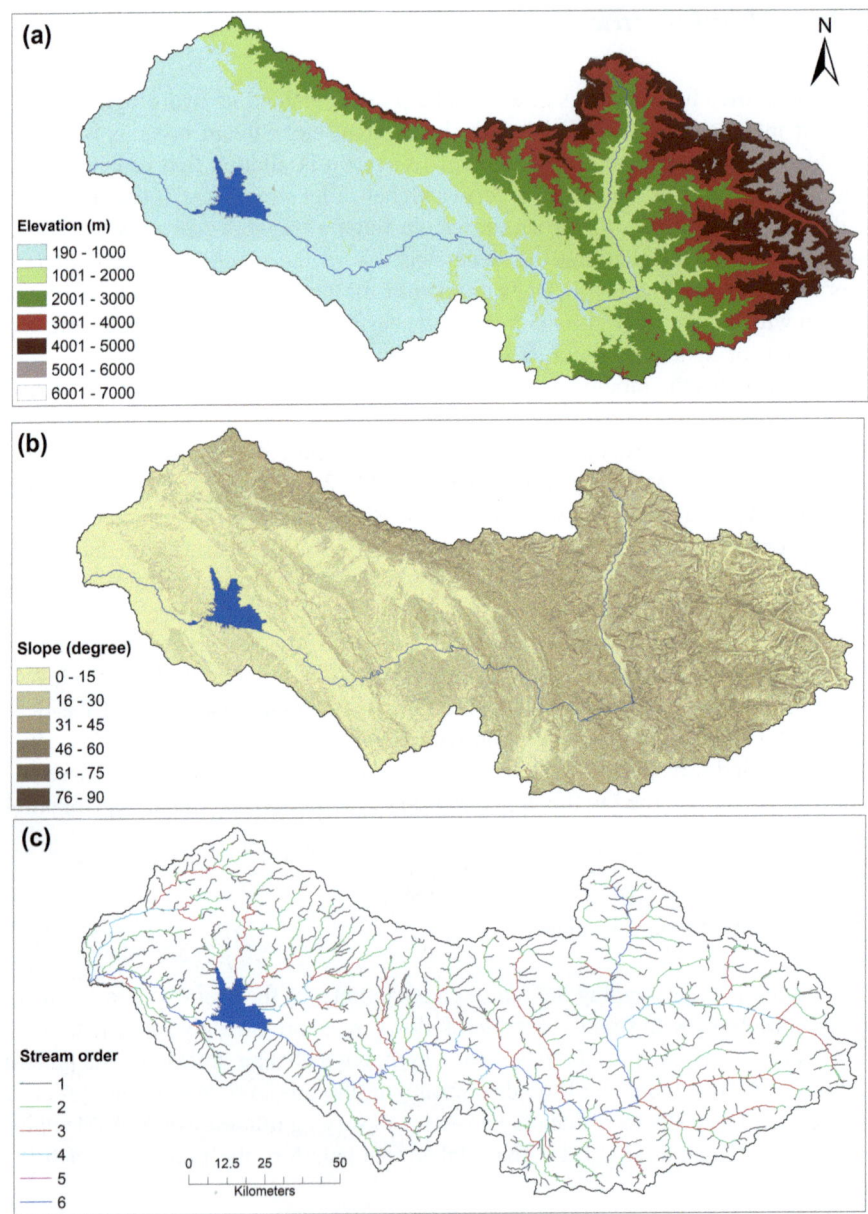

Fig. 2.11 Elevation, slope, and stream order of the Beas River basin

second-, third-, fourth-, fifth-, and sixth-order streams are 2576.4, 1366.3, 662.3, 242, 203, and 52.3 km, respectively (Table 2.1). The total stream length cumulative of all orders is 5102.3 km. The total stream length of the upper Beas basin is 2314.84 km with the contribution of first to sixth order being 1145.9, 606.58,

294.29, 153.81, 90.34, and 23.92 km, respectively (Vishwas 2021). The total stream length of the Parbati basin is 2751.51 km, having lengths of 1886.73, 481.03, 192.20, 98.30, 51.67, and 41.58 km of different orders from first to sixth, respectively (Khan et al. 2021).

2.3.1.2 Mean Stream Length (Lsm)

It displays the basin aspect by investigating the different components of the drainage network, and it has influenced on basin surface (Strahler 1964). The Lsm of the Beas basin of first- and second-order streams are 3.27 and 7.94, respectively (Table 2.1). The high value of Lsm indicates the landform's mature geomorphic development of the basin (Mahala 2020). The Lsm values of third-, fourth-, fifth-, and sixth-order streams are 16.98, 26.89, 101.48, and 52.27, respectively. The anomalies in Lsm indicated the change in slope and geological characteristics (Mahala 2020). The mean stream length of the upper Beas basin is 1.21, 4.33, 8.92, 21.97, 30.11, and 23.92 km (Vishwas 2021), and the Lsm of the Parbati basin is 0.70, 0.86, 1.64, 4.10, 10.33, and 41.58 km (Khan et al. 2021) of stream order from first to sixth, respectively.

2.3.1.3 Stream Length Ratio

The stream length ratio of the Beas basin starts with 2.43 for second to first order, 2.14 for third to second order, 1.58 for fourth to third order, 3.77 for fifth to fourth order, and 0.52 for fifth to sixth order, and the overall mean value for stream length ratio is 2.09 (Table 2.1). The stream length ratio for the Parbati basin varies from 0.25 to 0.80 (Khan et al. 2021).

2.3.1.4 Bifurcation Ratio (Bf)

High undulated terrain areas have higher values of Bf, viz., >5, 3 to 5 for moderately undulated terrain, and <3 for flat terrains. The Bf of the Beas basin from first order to fifth order ranged from 2 to 4.6 with a mean Bf of 4 (Table 2.1). The Bf values from first to fifth order are 4.6, 4.4, 4.3, 4.5, and 2, respectively. The inconsistency in the Bf indicates the inconsistency in the lithological adjustment of the basin (Mahala 2020). The mean Bf value of 2.17 indicates flat terrain (Asfaw and Workineh 2019). The mean Bf of the upper Beas basin is 4.21 with the highest Bf (6.76) of first to second order which indicates accelerated erosion (Vishwas 2021). The mean Bf of the Parbati basin is 4.85 with values ranging from 4.77 to 5.2 (Khan et al. 2021).

Table 2.2 Estimated values of morphometric aerial and relief parameters of the Beas basin

Areal aspects	Estimated value	Relief aspects	Estimated value
Drainage density	0.35	Absolute relief (m)	6188
Stream frequency	0.07	Relative relief (m)	5998
Texture ratio	1.25	Relief ratio	0.018
Circulatory ratio	0.28	Dissection index	0.96
Elongation ratio	0.62	Ruggedness index	2.06
Drainage texture	0.97		
Form factor	0.29		
Constant of channel maintenance	0.80		

2.3.1.5 Drainage Density (Dd)

The overall Dd of the Beas basin was observed to be 0.35 km/km^2 (Table 2.2). The low Dd indicates the basin character to be more permeable and with less runoff. Dd of 0.47 km/km^2 is observed in the Beas basin up to Aut (Vishwas 2021), and Dd of 1.55 km/km^2 is observed in the Parbati basin (Khan et al. 2021). The Dd in the upper sections of the river basins is observed to be more due to a weak impermeable surface facilitating less infiltration, high runoff, high relief, and sparse vegetation (Khan et al. 2021).

2.3.1.6 Stream Frequency (Fs)

The stream frequency of the Beas basin is 0.07 streams/km (Table 2.2). The low "Fs" value of the basin is due to the permeable surface and less runoff. The opposite of this is observed in the upper reaches of the basin due to high relief. The "Fs" of the Beas basin is 0.23 streams/km (Vishwas 2021), and for the Parbati basin, it is 1.91 (Khan et al. 2021), which is quite high and suggests high relief and poor infiltration.

2.3.1.7 Drainage Texture (Dt)

The Dt values of <2 indicate very coarse texture, 2–4 coarse texture, 4–6 moderately coarse, and 6–8 and >8 very fine drainage texture (Smith 1950). The Beas basin has a coarse texture with a value of 0.97 (Table 2.2). The values of "Dt" for the Beas basin are 2.92 (Vishwas 2021) and 11.8 for the Parbati basin (Khan et al. 2021). The value 2.92 denotes coarse texture, and 11.8 suggests the basin has a very fine texture resulting in very high Dd and Fs.

2.3.1.8 Circularity Ratio (Rc)

It is controlled by the geological structure, stream length, geomorphic character, stream frequency, soil properties, climate type, and land use/land cover in a basin (Rai et al. 2018). The circularity ratio of the Beas basin is 0.28 which implies an elongated shape of the basin (Table 2.2). An elongated shape is indicative of the fact that the region contains permeability and homogeneity in the surface geology (Asfaw and Workineh 2019). "Rc" for the Beas basin is 0.41, reflecting the strongly elongated shape of the basin (Vishwas 2021), and "Rc" is 0.24 for the Parbati basin which is also of elongated shape (Khan et al. 2021).

2.3.1.9 Elongation Ratio (Re)

"Re" values of <0.5 is tectonically active, 0.5–0.75 is slightly active, and >0.75 has been inferred as tectonically inactive (Sharma and Sarma 2013). "Re" value can be categorized as circular (0.9–0.10), oval (0.8–0.9), less elongated (0.7–0.8), elongated (0.5–0.7), and more elongated (<0.5) (Rai et al. 2018). The elongation ratio of the Beas basin is 0.62 indicating an elongated shape (Table 2.2). This value indicates steep slopes and high relief with an elongated shape of the basin (Strahler 1957). The Beas basin value is 0.78 characterized by steep slopes, high relief, and the presence of tectonic activity (Vishwas 2021), and the value for the Parbati basin is 0.06 which depicts the highly elongated shape of the basin (Khan et al. 2021).

2.3.1.10 Form Factor (Ff)

It is used to determine the geometry of basins, associated hydrological parameters, the pace of the erosional process, sediment transport, flood formation, and flooding corridor. It is the proportion of a basin's size to its square root of axial length (Horton 1932). The Beas basin's form factor is 0.29, which indicates that it is an elongated basin with low peak flow and takes longer to pass from a point, reducing the likelihood of flooding in the basin's lower portion (Table 2.2) (Sarkar et al. 2020). The "Ff" value for the Beas basin is 0.48, showing a less elongated shape (Vishwas 2021). The "Ff" value of the Parbati basin is 0.24, indicating a more elongated shape that tapers toward the northeast (Khan et al. 2021).

2.3.1.11 Constant of Channel Maintenance (CCM)

The Beas basin has a "CCM" value of 0.80 km^2/km which is low and indicates low infiltration and high runoff; water takes time to reach the main channel which results in high flood potential (Table 2.2) (Mahala 2020). The CCM value is 2.14 for the Beas basin and 0.64 sq. ft. for the Parbati basin, indicating a lower value and high flood potentiality (Khan et al. 2021).

2.3.1.12   Basin Relief (R, H)

The Basin relief includes absolute relief (R) and relative relief (H). Generally, basin relief is high on hilly terrain than in plain or plateau areas. The erosional character of any basin is perfectly indicated by the basin relief (Mahala 2020). The factors like slope steepness, sediment transport, flood probability, and river carrying capacity are controlled by basin relief (Hadley and Schumm 1961). The area of the Beas basin predominantly contains high peaks which shows the basin's susceptibility to erosion (Vishwas 2021). The highest relief or absolute relief (R) of the Beas basin is 6188 m, and the lowest is 190 m (Table 2.2). The basin relief or the relative relief (H) of the Beas basin is 5998 m.

2.3.1.13   Relief Ratio (Rr)

The Rr of the Beas basin (the area is 14,490 km^2) is 0.0189 which is very low (Table 2.2). The low "Rr," despite having high basin relief, is due to the large area of the basin (Vishwas 2021). The Rr for the Beas basin (area of 5300 km^2) is 0.056 (Vishwas 2021). A relief ratio of 0.06 is observed in the Parbati basin (area of 1773 km^2) (Khan et al. 2021).

2.3.1.14   Dissection Index (Di)

It is the measurement of vertical erosion occurring in a basin. A value near "0" indicates maximum denudational stages of evolution, which implies approximately all the stages have been attained and a value near "1" indicates minimum denudational stages of evolution, which implies all the stages have not been attained. Through "Di," the stage of geomorphic development can be identified. It is now clear that a high Di value is the characteristic property of mountain or elevated areas and a lower value is of plain areas. The Beas basin has a dissection value of 0.96 which shows the young geomorphic character of the basin (Table 2.2). The past study has computed the "Di" for the upper Beas basin as 0.86 (Vishwas 2021).

2.3.1.15   Ruggedness Index (Ri)

As a result of the absolute relief which seems to be high, the "Ri" of the Beas basin is high with a value of 2.06 (Table 2.2). Results show the long and steep slopes. The basin is in the early/young stage of geomorphic development. "Ri" for the upper Beas basin is 2.31 (Vishwas 2021) and 8.53 for the Parbati basin (Khan et al. 2021).

2.3.1.16   Slope (θ)

Slopes of the Beas basin are classified into three high categories (27°–82°), moderate (13°–27°), and low (0°–13°). A high-slope (27°–82°) degrees is observed on the upper course of the Beas basin around the main channel roughly up to Pandoh Dam; this high slope causes low infiltration and high runoff, making this region a potential place for making a dam (Table 2.2). The lower course around the mouth of the river has a low slope from 0° to 13°.

2.4   Climate

The Beas basin has a huge variation in climatic conditions due to variations in altitude (200–6500 m AMSL). Based on Koeppen's climatic classification, the study area falls in "Cwg" and "Cfa" categories which are warm temperate monsoon climates with dry winters and humid subtropical climates, respectively (Kumar and Rao 2021). The upper Beas basin, a hilly and mountainous region, experiences a pleasant and cool climate throughout the year with heavy snowfall during the winter months. However, the lower basin is a highly monotonous alluvial plain with extremely warm temperatures during the summers. Physiographically, it is semiarid and hot in plains, subhumid and less hot in valleys and foothills, warm and temperate in hilly and mountains, cool and temperate in middle Himalayas, and cold alpine and glacial in Lahaul and Spiti region of the Beas basin. The monthly mean air temperature of the basin is shown to increase by around 23 °C in June and July and decrease by less than 5 °C in January (Fig. 2.12). The variations in height and aspect

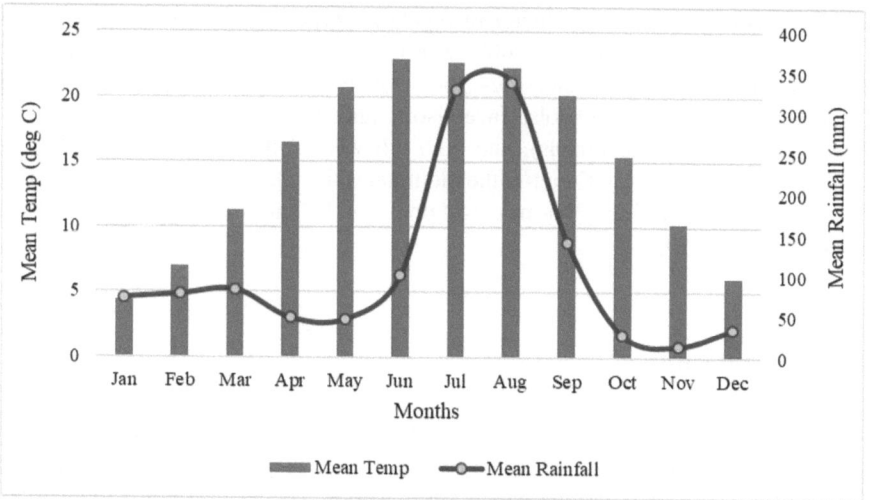

Fig. 2.12 Normal monthly mean air temperature and rainfall of the study area (1980–2020)

affect the weather. About 1314 mm of rainfall is recorded on an annual average in the basin. The area received the highest rain in July/August (338 mm) and the lowest in November (14 mm) (Fig. 2.12).

The basin experiences the following four seasons, depending on general climatic conditions, namely, winter season (December–March), pre-monsoon season (April–June), monsoon season (July–September), and post monsoon (October–November) (Figs. 2.13a, 2.13b and 2.14a, 2.14b) (Jain et al. 2009). A mid-latitude extratropical weather system moving eastward from the Caspian Sea is the cause of the precipitation during the winter season. Western disturbances are the name given to the winter weather system that approaches India from the west via Pakistan, Afghanistan, and Iran. In general, the Greater Himalayas receives snow during this time of year, and the middle Himalayas receives snow and rain, while the outer Himalayas and the nearby north Indian plains receive light to moderate rain. During the winter, there is intermittent precipitation. Pre-monsoon season, which typically lasts from April to June, is regarded as the transitional period between winter and the southwest monsoon. Air mass convective storms are primarily responsible for light to moderate rainfall. Convection intensifies as a result of this season's rising trend in temperature in the Himalayan region. During the monsoon season of the year, the moist air currents from the Bay of Bengal typically bring rain to the Himalayas, and the basin received heavy rainfall. Sometimes, in connection with specific weather conditions, both monsoon branches—the Bay of Bengal and the Arabian Sea—arrive simultaneously in this area, signaling the start of the monsoon season.

2.5 Soil

The soil in an area is a determinant of hydrological processes and landscape changes. Soil availability is a decisive factor in vegetation type and distribution. The Krishi Vigyan Kendra, Kullu district (2008–2009),[1] has broadly divided the soils of the Beas River basin into four categories: soil in valleys (Entisols, Inceptisols), mid-hill mild temperate areas (Entisols, Inceptisols, and Mollisols), high-hill temperate areas (Alfisols and Inceptisols), and high hill wet temperate areas. According to Fig. 2.15, the soil Magar-Keran-kathel dominates the majority of the research area, followed by Kippar-Urla, Dha-masan-Mamel, and Bitohi-Urla (SLUSI 2013; (Rani and Sreekesh 2019).

2.6 Vegetation

The river basin has deciduous forests in the lower portion and alpine flora in the higher portion due to a significant elevational change (Figs. 2.16 and 2.17a). Above the limit of tree growth are the Alpine scrub woods. The basin is home to a variety

---

[1] Zonal Project Directorate: http://zpdzone1.org/zone1/download.php?file=Kullu0882.doc

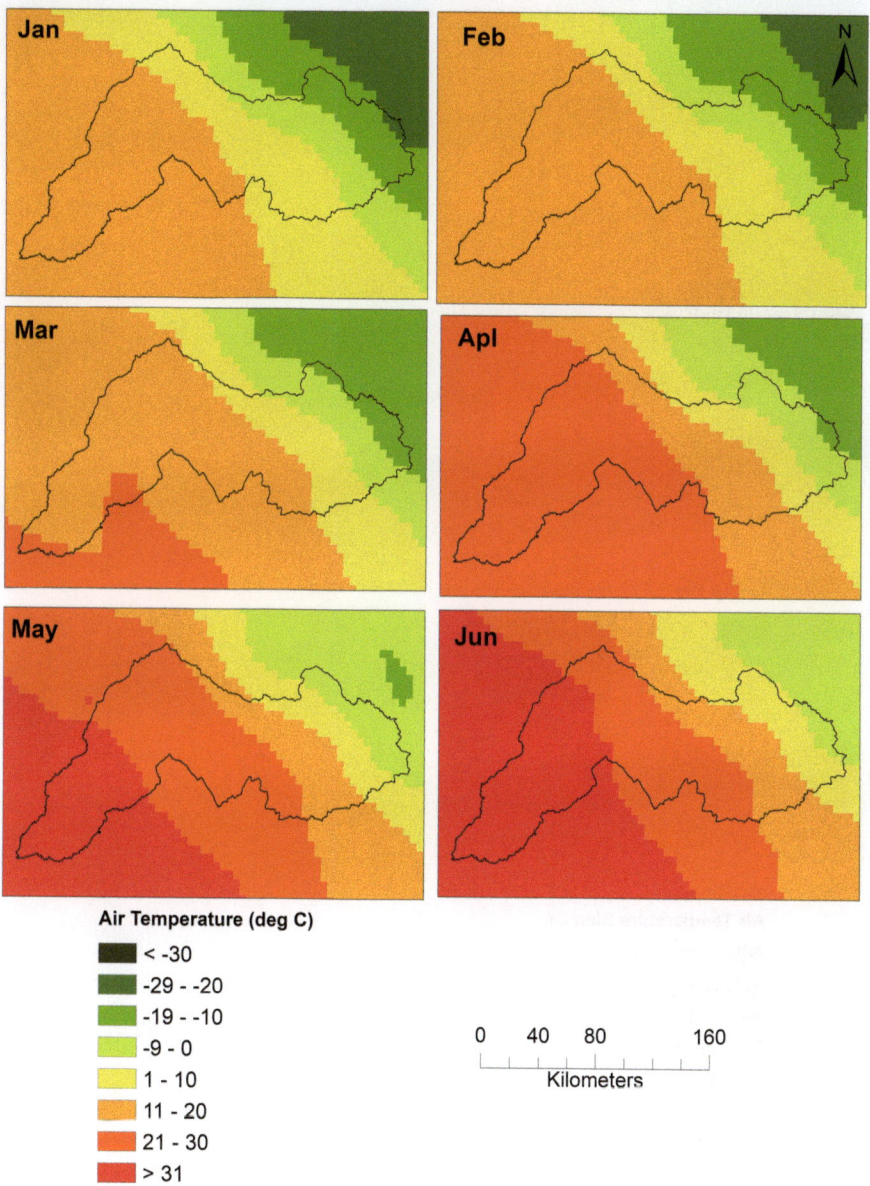

Fig. 2.13a Distribution of mean air temperature of the study area (1980–2020)

of medicinal plants, including aconite, dhoop, and karru. The most valuable wood forest in the watershed is the moist Deodar Forest. Pure spruce, pure silver fir, silver fir spruce, and spruce-deodar formation make up the mixed coniferous forest. These appear between 2000 and 3000 meters above the Deodar and Kail zone. Between 2000 and 3000 m, in moist depressions frequently along the Nalas, is where the

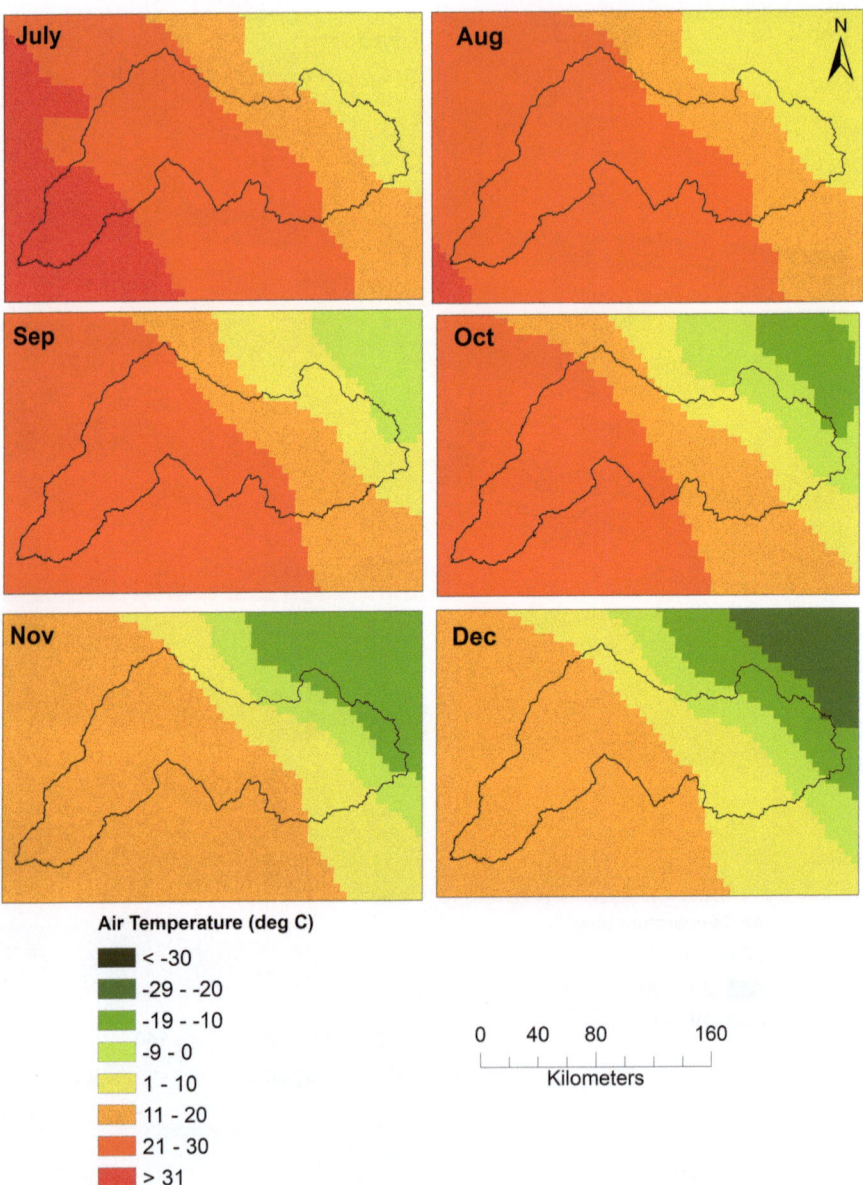

Fig. 2.13b Distribution of mean air temperature of the study area (1980–2020)

moist temperate deciduous forest may be found. The primary tree species found in quartzite rocks include chestnut, walnut, maple, oak, and poplar. Over the majority of the altitudinal range, devdar and karsu are mostly found on south-facing slopes, whilst fir and spruce are primarily connected with north-facing slopes (Fig. 2.17a). The NDVI values show seasonal variations in the extent of vegetation in the basin

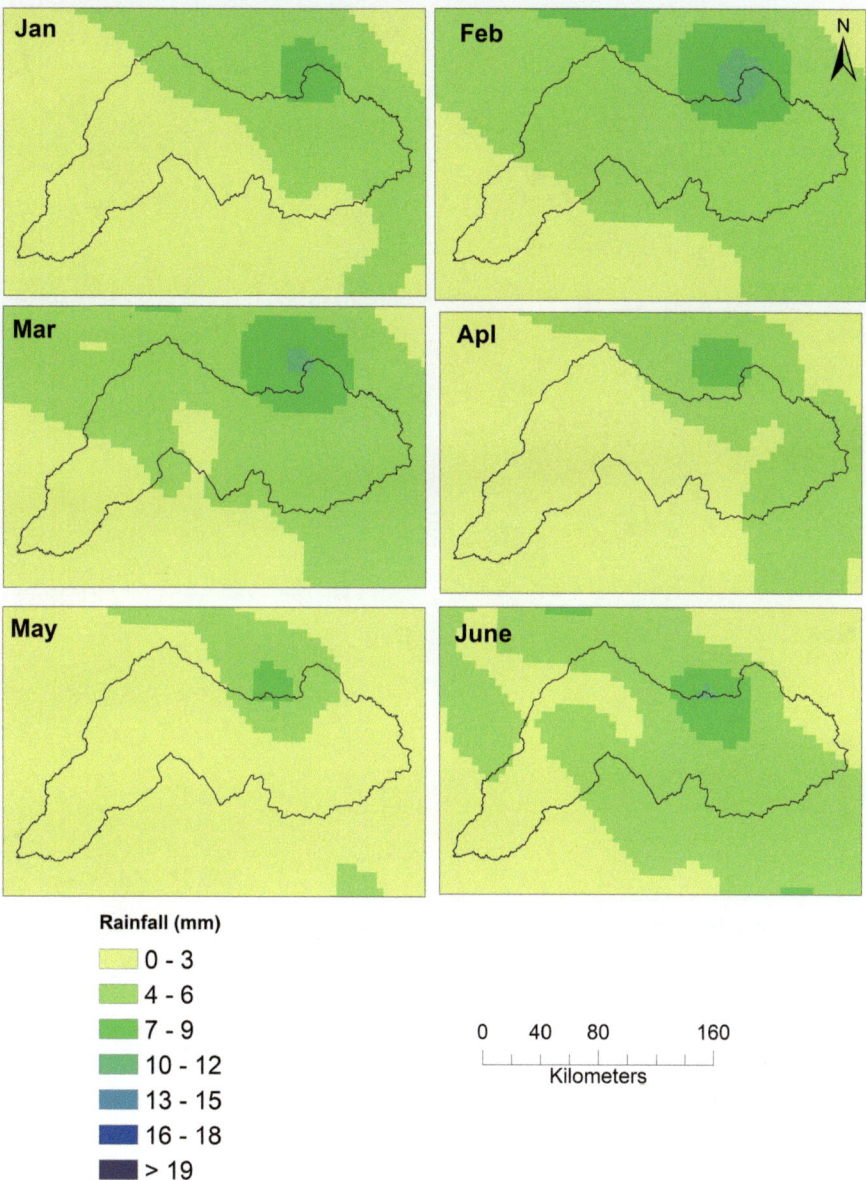

Fig. 2.14a Distribution of rainfall of the study area (1980–2020)

(Figs. 2.17a and 2.17b). Monsoon season has high values of NDVI. It declines with height, and grasslands become prominent nearby snow areas. Additionally, forests help to keep the soil in place on a steep slope, preventing both soil erosion and other disastrous disasters. Overgrazing and the lack of continual afforestation are the primary drivers of the overall reduction in forest cover.

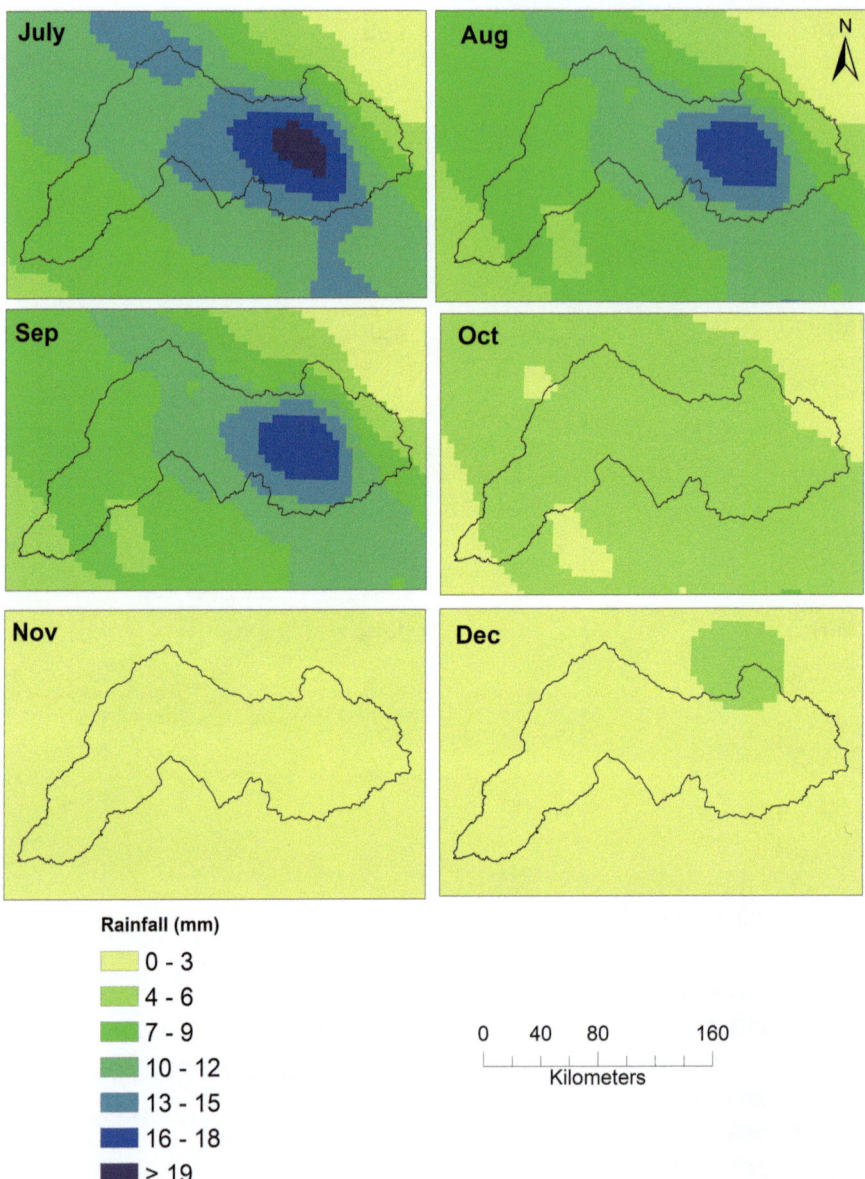

Fig. 2.14b Distribution of rainfall of the study area (1980–2020)

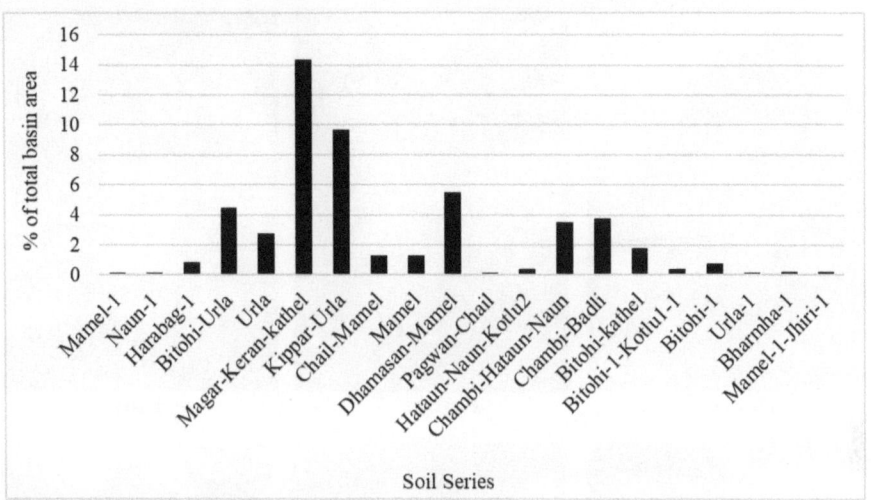

Fig. 2.15 Soil series distribution in the study area

2.7   Land Use/Land Cover

Detailed land use/land cover (LULC) analysis of the basin is given in Chap. 4. The study region is mostly covered by forest, with smaller areas of barren, uncultivated, and wasteland (BUW); snow and glaciers; grassland; cultivated land; and settlements. The north, northeast, and eastern portions of the study area are covered with snow and glaciers, whereas the center of the study area is covered in forest. Due to the considerable altitudinal differences, settlements are concentrated in the valleys and cover a smaller area. A select few towns, including Manali, Nagar, Kullu, Bhuntar, Kasol, and Banjar, are situated in the upper Beas River basin.

2.8   Demography

The Beas River flows through 15 districts of Himachal Pradesh (Kullu, Mandi, Chamba, Una, Hamirpur, Kinnaur, Lahaul and Spiti, Shimla, and Kangra) and Punjab states (Amritsar, Firozpur, Gurdaspur, Hoshiarpur, Jalandhar, and Kapurthala) (Fig. 2.18). Major towns in the basin are Manali, Naggar, Katrai, Raison, Kullu, Bajura, Pandoh, Mandi, Jogindernagar, Sujanpur Tihra, Nadaun, Dehra Gopipur, Kangra, Dharamshala, Chamunda, Palampur, Jwalaji, and Chintpurni. Kinnaur, Lahaul and Spiti, and Shimla share a very small area of the total basin. Its catchment area has 3 large cities, 44 towns, 4331 settlements, 19 tourist stations, and 2 airports.

The basin supports a large population of northern India; mainly the lower parts are much denser. The lower basin falling in Punjab state had a relatively dense

Fig. 2.16 Important trees in the high altitudes of the study area (based on a 2017 survey)

population owing to its supportive relief, agricultural intensification, and economic development as compared to rugged, hilly, and inaccessible regions of the upper basin. The basin serves around 13.8 million population in the state of Punjab and Himachal Pradesh (Census of India 2011a, b, c, d, e, f, g, h, i, j, k, l). District-wise total and rural/urban population of the Beas basin is given in Fig. 2.18. Regarding population, Kangra district is ranked first among the districts. The Hamirpur district in the state has the highest population density, with 407 people/km². About 7.9% of all people living in the Punjab state are in the Jalandhar district. With 80 people/ km², the Chamba district has the ninth-highest population density in the state and ranks seventh among the districts in terms of population (Fig. 2.18). In comparison to the state's decadal population increase of 12.9%, the Una district saw a 16.3% decadal population rise from 2001 to 2011. In the basin, around 80% of the overall population lives in rural areas on average. In the Himachal Pradesh basin, the rural population is greater than 90%. Out of the total population of Jalandhar district,

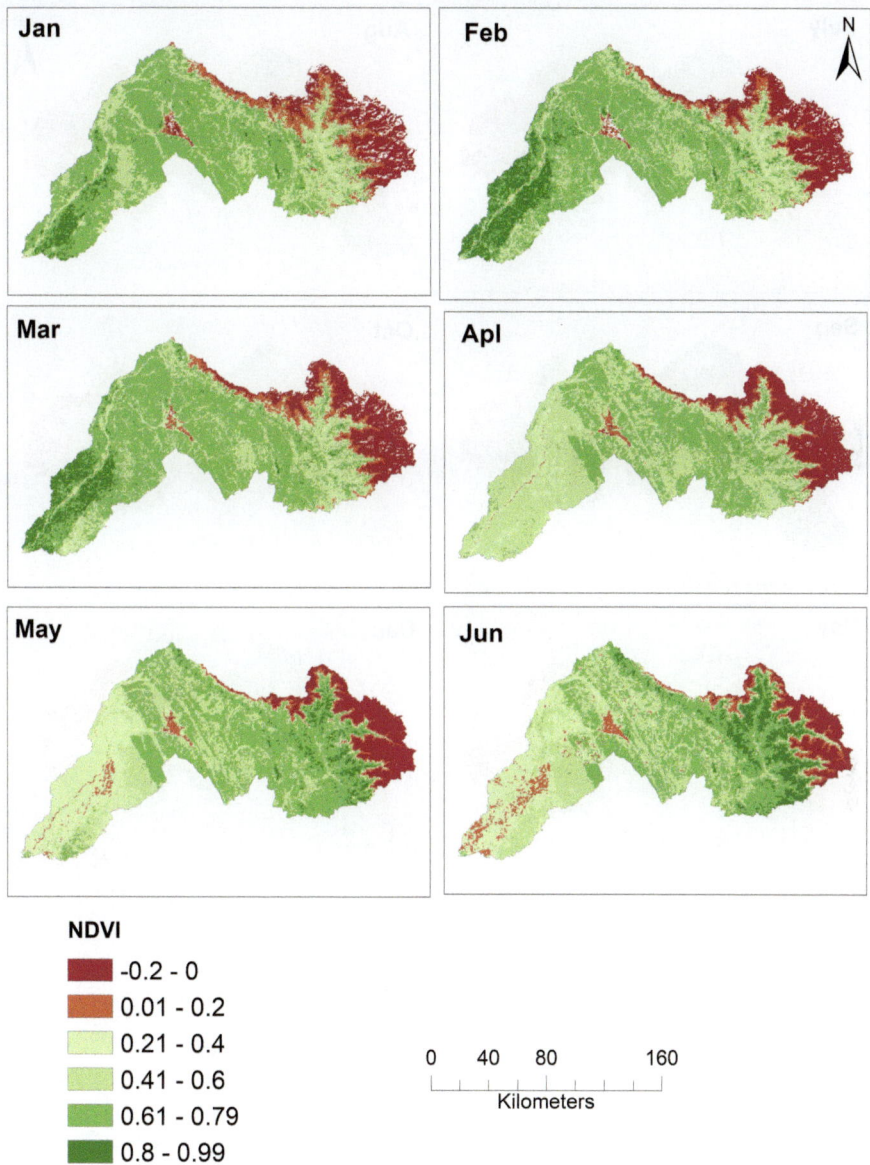

Fig. 2.17a Monthly variations in NDVI of the study area

47.1% is in rural areas and 52.9% is in urban. According to the Census of India (2011g, j), Amritsar recorded the highest population density with 956 persons/km^2 followed by Jalandhar (861 persons/km^2). On the other hand, the upper parts of the basin like Lahaul and Spiti and Kinnaur are the most sparsely populated with a population density of only 2 and 13 persons/km^2, respectively. Tourists travel from

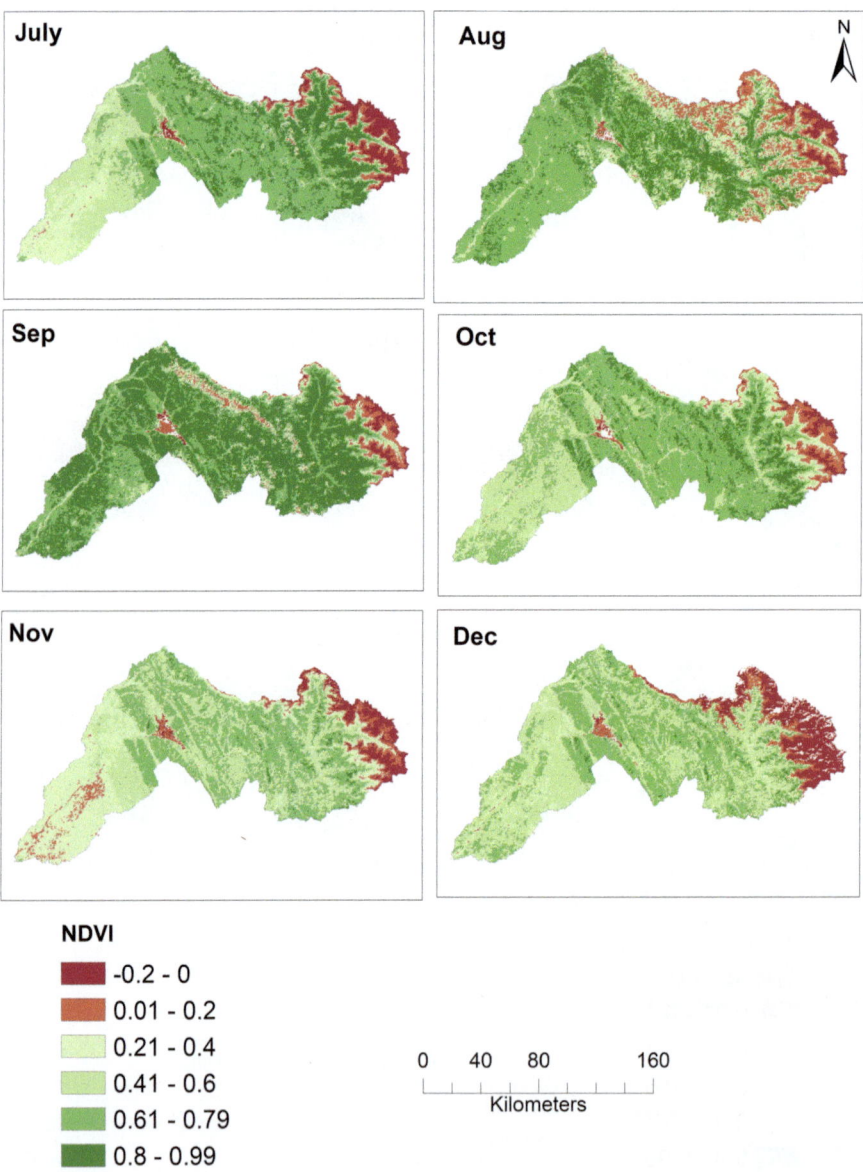

Fig. 2.17b Monthly variations in NDVI of the study area

all over the world to Manali, and Bhuntar is a significant city as well. Manali and Bhuntar witnessed growth rates of 252% and 62%, respectively, between 1981 and 2011, according to the Census of India (2011). Manali saw an unusual growth rate of 157.50% from 1991 to 2001, which was greater than that of the urban centers in Himachal Pradesh. As a result, Manali drew a sizable population during this time. Bhuntar saw a rapid population increase of 43.3% within the same 10 years.

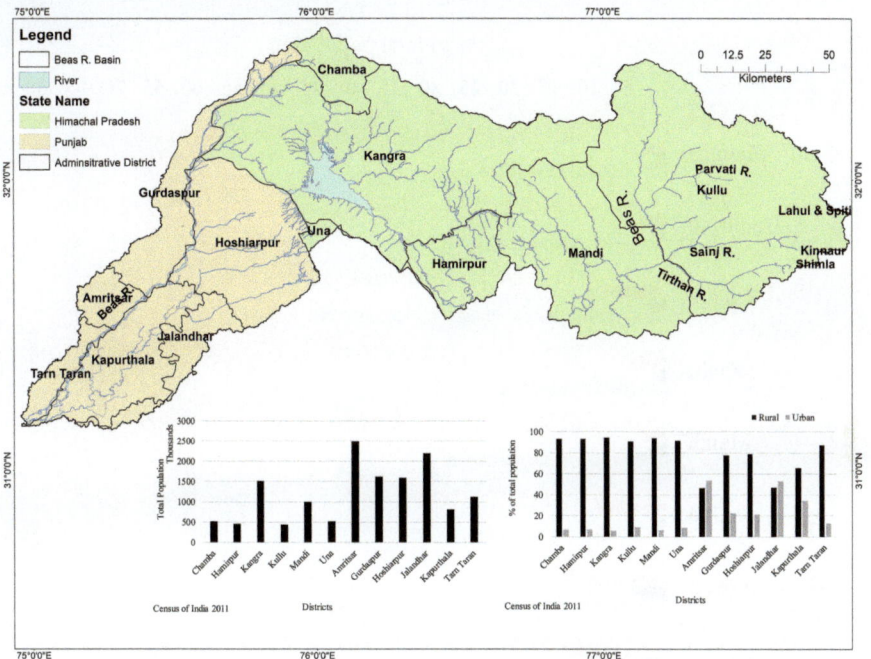

Fig. 2.18 Administrative division with total, rural and urban population of the Beas River basin

2.9   Economy

The economy of the basin depends on agriculture (including horticulture), industries, and tourism. On average, 49% of the total population of the basin are cultivators and agricultural laborers, while only 3% are in household industries (Figs. 2.19
and 2.20). In the Himalayan part (Himachal Pradesh) of the basin, more than 60%
populations are cultivators and agricultural laborers, while 2% are in household
industries. The data showed the dominance of agriculture in the area. Punjab forms
part of the wheat rice belt of India where the green revolution was implemented. It
is also the food basket of India. The economy of the Chamba district mainly depends
on agriculture and holds the fifth position among the districts of the state. Kangra
and Hamirpur hold the second and seventh positions in terms of cultivators among
the districts of the state, respectively. According to the district's employees in
Amritsar, 13.58% are cultivators, 13.25% are agricultural laborers, 4.5% work in
the domestic industry, and 68.64% are other workers, meaning that 26.83% are
employed in agriculture and 73.18% in nonagricultural sectors. The people of
Chamba used to undertake various activities like metal craft, embroidery, wood
carving, stone carving and painting, etc. The stunning traditional weaving handi-
crafts produced in the Kullu district are a source of great pride. The area is well-
known for its colorful Kullu hats, shawls, pattoos, and patties (local tweed). The

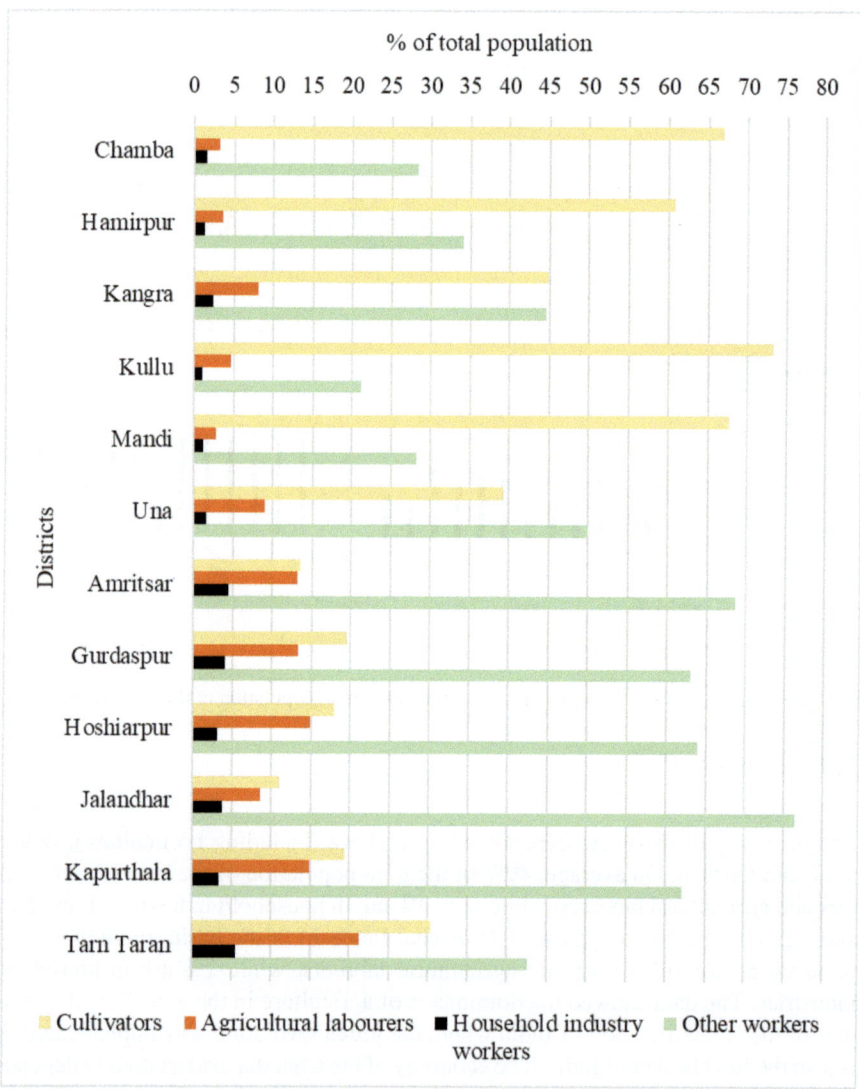

Fig. 2.19 District-wise percentage of working population in the Beas basin. (Census of India 2011)

Amritsar area is a leading producer of woolen textiles such as worsted, tweeds, blankets, and shawls in the nation.

The most significant economic activity in the basin is also tourism, especially in the higher reaches. With well-known tourist destinations including Chamba town, Khajjiar, Dalhousie, and Bharmour, Chamba district has a significant position on the state's tourism map. The Hamirpur district is renowned for having appealing tourist attractions. The district's well-known tourist attractions are Hamirpur, Tira Sujanpur, Nadaun, and the Temple of Deothsidh Baba Balak Nath. The Kangra area

Fig. 2.20 Agriculture in the Beas River basin (fieldwork 2016, 2017, 2022): (**a**) apple plantation around Manali, (**b**) wheat, (**c**) sheep husbandry, and (**d**) pomegranate plantation around Bhuntar

is renowned for having appealing tourist attractions. The district's well-known tourist destinations include Dharamsala, Kangra, Palampur, Baijnath, Jawalamukhi, and Masroor. The Kullu area is renowned for having appealing tourist attractions. Manikaran, Nagar, Kullu, and Manali are well-known tourist destinations in the region (Fig. 2.21). As each of its villages has a god or goddess, known as Deo, Devta, or Devi, Kullu Valley is also known as the "land of gods" or "Dev Bhumi." The well-known Hadimba Devi Temple is located near Manali. Along with well-known religious destinations like Chintpurni, Dera Baba Barbhag Singh, Dera Baba Rudru, and Gurudwara Baba Sahib Singh Ji Bedi, Una district is significant on the state's religious tourist map.

2.9.1 Hydroelectric Projects

The 51 hydroelectric projects spread out across the Beas basin in Himachal Pradesh have a combined capacity of 4877.70 MW (for projects larger than 5 MW). Out of these 51 projects, 22 are completed (with a total installed capacity of 2820.90 MW), 5 are being built (with a total installed capacity of 947 MW), 20 are being

Fig. 2.21 Tourists' places in the Beas River basin (fieldwork 2017, 2022): (**a**) base camp for the Beas Kund trek, (**b**) Solang valley. (**c**) Manali main market, (**d**) Government forest park, Manali, (**e**) Hadimba Devi Temple at Manali, (**f**) Naggar Castle, (**g**) Parvati valley, (**h**) Sainj valley

investigated (with a total installed capacity of 1028.90 MW), and 4 are still awaiting allocation (Ministry of Environment, Forest and Climate Change Government of India 2019). There are about 11 dams (7 in Himachal Pradesh and 4 in Punjab) and 7 barrages/weirs/annicuts on the Beas River including 16 powerhouses. There are many small- and large-sized hydropower projects in the upper Beas basin (Table 2.1; Figs. 2.22 and 2.23). Prominent dams of the upper Beas basin in Himachal Pradesh

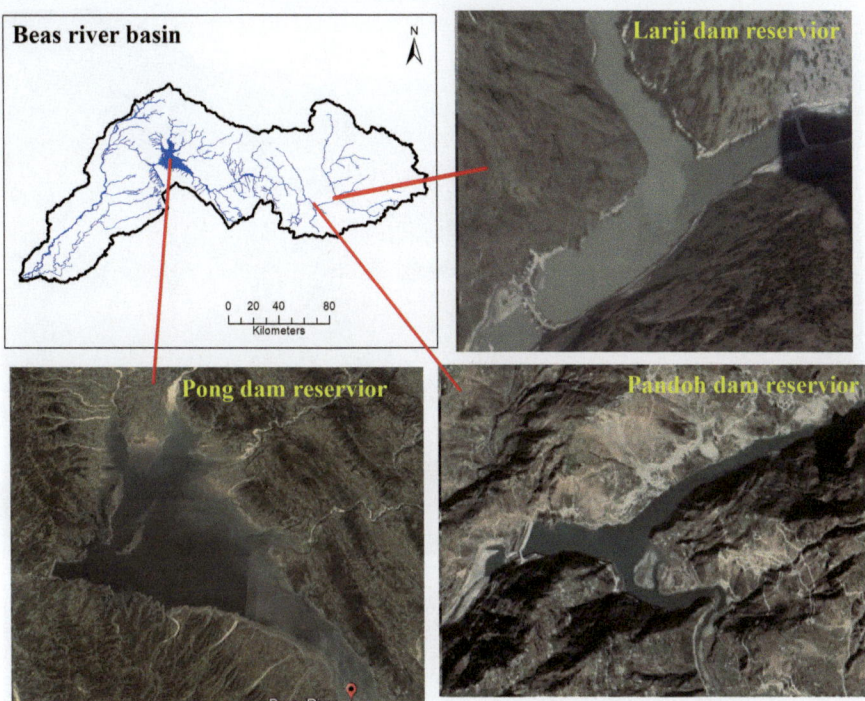

Fig. 2.22 Important reservoirs of the Beas River basin (prepared from Google Earth)

are the Pong Dam on the Beas River (district Kangra), Malana II Dam on Malana Nallah (Kullu district), Bassi Dam on the Beas River (Mandi district), Parbati II Dam on Parbati River (Kullu district), Parbati III Dam on Sainj River (Kullu district), Larji Dam on Beas River (Mandi district), and Pandoh Dam on Beas River (Mandi district). In the lower Beas basin, all the dams are located in the Hoshiarpur district of Punjab including Janauri Dam on Janauri Khad River, Dholbaha Dam on the Dholbaha Khad River, Sal Dam on Damsal River, and Thana Dam on Khwaja River (Sharma and Paithankar 2014) (Table 2.3).

It has major irrigation projects to provide water to agricultural fields. Some important names are Bist Doab Canal System, Shah Nehar Canal System, and Upper Bari Doab Canal (UBDC) System (Table 2.4).

2.10 Flood Events

River sections in the Beas basin's north and east are prone to flooding, which is primarily controlled by intense rainfall and high runoff characteristics. Bahang Manali (Beas), Kullu-Bhuntar (Beas), and Manikaran-Kheer-Ganga (Parvati) river segments have been divided into extremely high and high fast-food zones (Singh

Fig. 2.23 Reservoir of the (**a**) Pandoh and (**b**) Larji Dam (fieldwork May–June 2017 and 2022)

et al. 2021). Thirty-one flash food events occurred in Himachal Pradesh from 1990 to 2018 (Singh et al. 2021). Out of these, the majority of the events occurred in the Beas River basin. The flood situation deteriorated between 1992 and 1996 when Kullu district's Beas River was hit by a flood (Chandel et al. 2014). Extreme flash floods that wiped away vast stretches of arable land and crops were reported in 2000, 2001, 2005, and 2007. The Beas River section between Manali and Kullu had the worst flooding from 2011 to 2018. In the basin, the last flood occurred in 2018,

Table 2.3 Hydro electric projects on the Beas River

Dam	Mega Watt (MW) to be generated	Operational year	Type
Pandoh Dam	990	1997	Reservoir
Larji Dam	126	2006	Reservoir
Malana I	86	2001	Runoff river
Allain Duhangan[a]	192	2010	Runoff river
Malana II	100	2010	Runoff river
Beas Kund	9	2012	Runoff river
Aleo-Manali	3	2009	Runoff river
Sainj	100	Under construction	
Parbati I	750	Under construction	
Parbati II	800	Under construction	
Parbati III	520	Under construction	
Bassi Dam	60	1981	Runoff river
Pong Dam	396	1975	Reservoir
Janauri Dam			
Damsal Dam		2001	Reservoir
Thana Dam	191		
Dholbaha Dam		1986	Reservoir
Pong Dam	396	1974	Reservoir

Source: Bhakra Beas management board (http://www.bbmb.gov.in/) [a]http://hppcblivedata.com/aleo_manali_hydro/

Table 2.4 Some major medium irrigation projects of the Beas River Basin

S.no.	Name	Type	Area (ha)
1	Shahnehar Irrigation Project	Major	74.82
2	Bist Doab Canal System	Major	748.32
3	Shah Nehar Canal System	Medium	744.31
4	Balh Valley II	Medium	24.08
5	Dholaba Dam	Medium	54.58
6	Balh Valley III	Medium	6.17
7	Sidhata Irrigation Project	Medium	38.54
8	Balh Valley I	Medium	33.66
9	Upper Bari Doab Canal (UBDC) System	Major	1473.18

which saw more rainfall (+49%) than usual during the monsoon season (June to September) of 2018. Significant damage was seen during the rainstorms on August 13 and September 23 and 24 along with a shift in the river course (Figs. 2.24a and 2.24b) (Government of Himachal Pradesh 2018). On August 13, 2018, September 23, and September 24, 2018, the Kullu district saw heavy rainfall that was estimated to be 920%, 3232%, and 362% over average, respectively. It leads to enormous loss of public and private property.

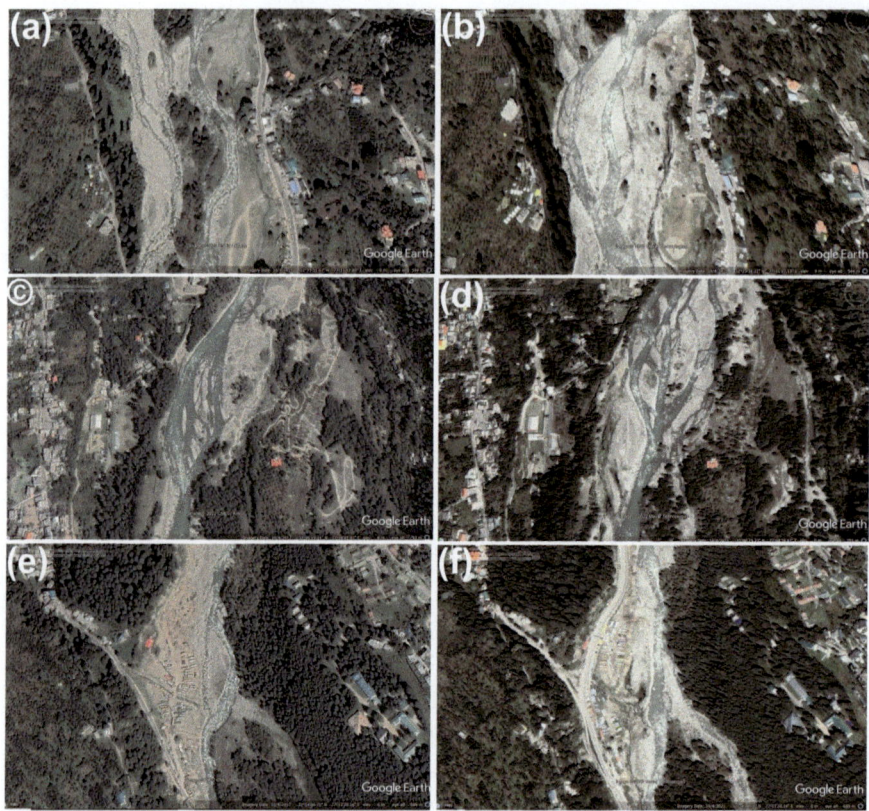

Fig. 2.24a Distribution of damages in the basin during 2018 flood. Left column shows pre-flood (October 4, 2017), and right column shows post-flood (October 4, 2021). (**a**, **b**) Near the Old Manal, (**c**, **d**) near Mahili, and (**e**, **f**) near Aleo

2.11   New Development

New hotels are coming along the banks of the river, particularly in the high reaches of the river for tourists. The Kiratpur-Ner Chowk Highway is developed from the section of NH-21 to four lanes in Himachal Pradesh. The 48-km Kiratpur-Nerchowk 4-lane project, which connects Manali with 5 tunnels and 22 Sutlej bridges, is still being built. The time it takes to get from Kiratpur to Kullu will be cut by up to 4 h because of this initiative. This will increase tourism and make people wealthier. It is modifying the landscape in the Beas basin and increasing the slope instability at a few locations. It is constructed along the river system and resulted in the banks cutting. There is a need to explore the environmental impact of the ongoing highway project on hydrological system dynamics.

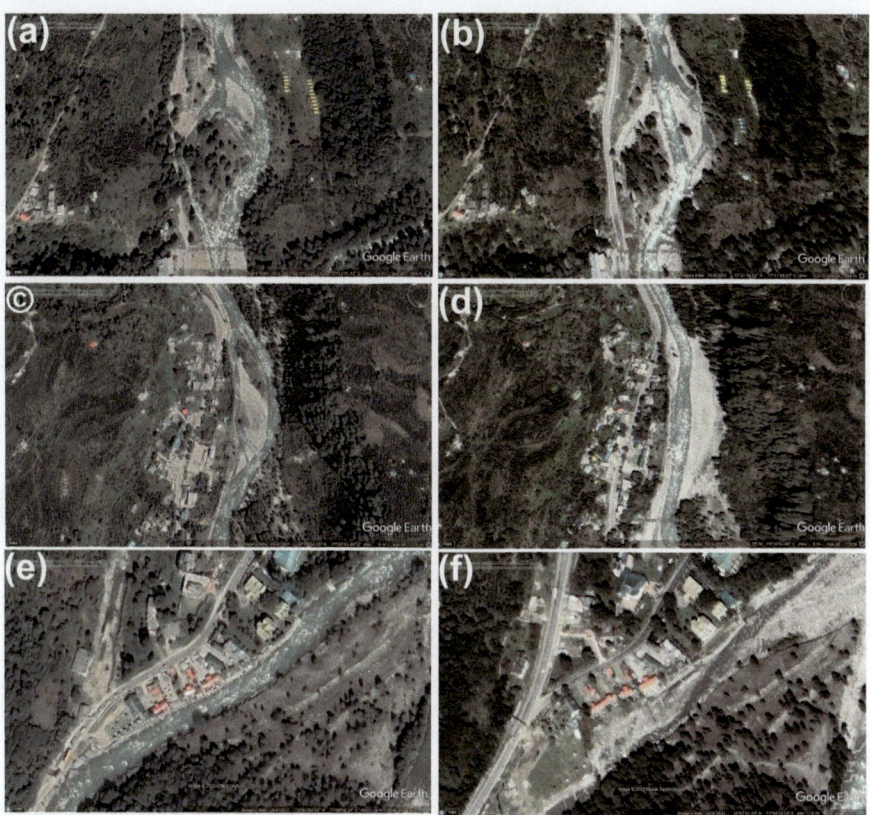

Fig. 2.24b Distribution of damages in the basin during the 2018 flood. The left panel shows pre-flood (October 4, 2017), and the right column panel shows post-flood (October 4, 2021). (**a, b**) Near Sajla, (**c, d**) near Karjan, and (**e, f**) near 15 miles

2.12 Summary

The Beas River is a dynamic river system that has diversity in physiography, climate, soil, and vegetation. It provides livelihood to lakhs of people upstream and downstream. The study of hydrological behavior is important for the population. The level of homogeneity is high in the lower basin compared to high-altitude areas of the basin. The basin has issues of landslide, avalanche, and flood which needs a detailed understanding of the basin. There is a need to understand the available conditions of the basin to estimate the situation in the coming times.

References

Asfaw D, Workineh G (2019) Quantitative analysis of morphometry on Ribb and Gumara watersheds: implications for soil and water conservation. Int Soil Water Conserv Res 7(2):150–157

Census of India (2011a) Himachal Pradesh, Series-03 Part XII-B, District Census Handbook, Chamba, Village and Town Wise, Primary Census Abstract (PCA), Directorate of Census Operations, Himachal Pradesh. Available via https://censusindia.gov.in/2011census/dchb/0201_PART_B_DCHB_CHAMBA.pdf

Census of India (2011b) Himachal Pradesh, Series-03 Part XII-B, District Census Handbook, Hamirpur, Village and Town Wise, Primary Census Abstract (PCA), Directorate of Census Operations, Himachal Pradesh. Available via https://censusindia.gov.in/2011census/dchb/0206_PART_B_DCHB_HAMIRPUR.pdf

Census of India (2011c) Himachal Pradesh, Series-03 Part XII-B, District Census Handbook, Kangra, Village and Town Wise, Primary Census Abstract (PCA), Directorate of Census Operations, Himachal Pradesh. Available via https://censusindia.gov.in/2011census/dchb/0202_PART_B_DCHB_KANGRA.pdf

Census of India (2011d) Himachal Pradesh, Series-03 Part XII-B, District Census Handbook, Kullu, Village and Town Wise, Primary Census Abstract (PCA), Directorate of Census Operations, Himachal Pradesh. Available via https://www.censusindia.gov.in/2011census/dchb/0204_PART_B_DCHB_KULLU.pdf

Census of India (2011e) Himachal Pradesh, Series-03 Part XII-B, District Census Handbook, Mandi, Village and Town Wise, Primary Census Abstract (PCA), Directorate of Census Operations, Himachal Pradesh. Available via https://censusindia.gov.in/2011census/dchb/0205_PART_B_DCHB_MANDI.pdf

Census of India (2011f) Himachal Pradesh, Series-03 Part XII-B, District Census Handbook, Una, Village and Town Wise, Primary Census Abstract (PCA), Directorate of Census Operations, Himachal Pradesh. Available via https://censusindia.gov.in/2011census/dchb/0207_PART_B_DCHB_UNA.pdf

Census of India (2011g) Punjab Series-04, Part XII-A, District Census Handbook Amritsar, Village and Town Directory, Directorate of Census Operations. Punjab. Available via https://censusindia.gov.in/2011ce sus/dchb/DCHB_A/03/0315_PART_A_DCHB_AMRITSAR.pdf

Census of India (2011h) Punjab Series-04, Part XII-B, District Census Handbook, Gurdaspur, Village and Town Directory. Directorate of Census Operations, Punjab. Available via https://censusindia.gov.in/2011census/dchb/0301_PART_B_DCHB_GURDASPUR.pdf

Census of India (2011i) Punjab Series-04, Part XII-B, District Census Handbook, Hoshiarpur, Village and Town Directory. Directorate of Census Operations, Punjab. Available via https://censusindia.gov.in/2011census/dchb/DCHB_A/03/0304_PART_A_DCHB_HOSHIARPUR.pdf

Census of India (2011j) Punjab Series-04, Part XII-B, District Census Handbook, Jalandhar, Village and Town Directory. Directorate of Census Operations, Punjab. Available via https://censusindia.gov.in/2011census/dchb/0303_PART_B_DCHB_JALANDHAR.pdf

Census of India (2011k) Punjab Series-04, Part XII-B, District Census Handbook, Kapurthala, Village and Town Directory. Directorate of Census Operations, Punjab. Available via https://www.censusindia.gov.in/2011census/dchb/0302_PART_B_DCHB_KAPURTHALA.pdf

Census of India (2011l) Punjab Series-04, Part XII-B, District Census Handbook, Tarn Taran, Village and Town Directory. Directorate of Census Operations, Punjab. Available via https://www.censusindia.gov.in/2011census/DCHB/0316_PART_B_DCHB_TARN%20TARAN.pdf

Chandel VB, Kahlon S, Brar KK (2014) Flood disaster in mountain environment: a study of Himachal Pradesh, India. Managing our resources: perspectives and planning. Bharti Publications, New Delhi, pp 11–21

Geological Survey of India (2012) Briefing book. Northern Region, Lucknow. https://www.gsi.gov.in/webcenter/

Government of Himachal Pradesh (2018) Revenue Department memorandum of damages due to flash floods, cloudbursts and landslides during the monsoon season – 2018. Disaster Management

Cell, HP Secretariat Shimla −2 Available via https://hpsdma.nic.in/WriteReadData/LINKS/c237c1ce-1102-4dce-853a-3472e83bed19.pdf

Hadley RF, Schumm SA (1961) Sediment sources and drainage basin characteristics in upper Cheyenne River basin. USGS water supply paper. US Geological Survey

Horton RE (1932) Drainage-basin characteristics. Trans Am Geophys Union 13(1):350–361. https://doi.org/10.1029/TR013i001p00350

Jain SK, Goswami A, Saraf AK (2009) Role of elevation and aspect in snow distribution in Western Himalaya. Water Resour Manag 23(1):71–83

Khan I, Bali R, Agarwal KK, Kumar D, Singh SK (2021) Morphometric analysis of Parvati Basin, NW Himalaya: a remote sensing and GIS based approach. J Geol Soc India 97(2):165–172

Kumar S, Rao KN (2021) Analysis of temperature and precipitation for assessment of climate change during 20th century in Beas River basin. Mausam 72(2):489–501

Kumar V, Singh P, Singh V (2007) Snow and glacier melt contribution in the Beas River at Pandoh Dam, Himachal Pradesh, India. Hydrol Sci J 52(2):376–388

Mahala A (2020) The significance of morphometric analysis to understand the hydrological and morphological characteristics in two different morpho-climatic settings. Appl Water Sci 10(1):1–6

Ministry of Environment, Forest and Climate Change Government of India (2019) Cumulative impact & carrying capacity study (CIA&CCS) of Beas sub basin in Himachal Pradesh. Available via http://www.indiaenvironmentportal.org.in/files/file/Report-CIAampCCS-of-Beas-sub-basin-in-HP.pdf

Momblanch A, Papadimitriou L, Jain SK, Kulkarni A, Ojha CS, Adeloye AJ, Holman IP (2019) Untangling the water-food-energy-environment nexus for global change adaptation in a complex Himalayan water resource system. Sci Total Environ 655:35–47

Rai PK, Chandel RS, Mishra VN, Singh P (2018) Hydrological inferences through morphometric analysis of lower Kosi river basin of India for water resource management based on remote sensing data. Appl Water Sci 8(1):1–6

Rani S, Sreekesh S (2019) Evaluating the responses of streamflow under future climate change scenarios in a Western Indian Himalaya Watershed. Environ Process 6(1):155–174

Sarkar D, Mondal P, Sutradhar S, Sarkar P (2020) Morphometric analysis using SRTM-DEM and GIS of Nagar river basin, Indo-Bangladesh Barind tract. J Indian Soc Remote Sens 48(4):597–614. https://doi.org/10.1007/s12524-020-01106-7

Sharma JR, Paithankar Y (2014) Indus Basin – Version 2.0, Department of Space, NRSC-ISRO. Central Water Commission, Ministry of Water Resources, New Delhi

Sharma S, Sarma JN (2013) Drainage analysis in a part of the Brahmaputra Valley in Sivasagar District, Assam, India, to detect the role of neotectonic activity. J Indian Soc Remote Sens 41(4):895–904

Singh S, Dhote PR, Thakur PK, Chouksey A, Aggarwal SP (2021) Identification of flash-floods-prone river reaches in Beas river basin using GIS-based multi-criteria technique: validation using field and satellite observations. Nat Hazards 105(3):2431–2453

SLUSI (2013) Soil resource mapping District Kullu and Mandi, Himachal Pradesh, Soil and Land Use of Survey India

Smith KG (1950) Standards for grading texture of erosional topography. Am J Sci 248(9):655–668

Strahler AN (1957) Quantitative analysis of watershed geomorphology. Trans Am Geophys Union 38(6):913–920. https://doi.org/10.1029/TR038i006p00913

Strahler AN (1964) Quantitative geomorphology of drainage basin and channel networks. In: Handbook of applied hydrology. McGraw-Hill, New York

Vercruysse K, Grabowski RC (2021) Human impact on river planform within the context of multi-timescale river channel dynamics in a Himalayan river system. Geomorphology 381:107659

Vishwas P (2021) Quantitative evaluation of drainage attributes to infer hydrologic and morphological characteristics of upper Beas Basin, Himachal Pradesh: a GIS-based approach. Geol Ecol Landsc 1–6. https://doi.org/10.1080/24749508.2021.1952766

Chapter 3
Climate Variability Assessment

Abstract The hydrology of a basin is mostly influenced by climate dynamics. Further global warming may exacerbate the effects of the water cycle, altered precipitation patterns, melting glaciers and ice sheets, shifting snow cover, rising mean sea level, and other changes. The purpose of this chapter is to describe the trends and relationships of mean air temperature (T_{mean}), evapotranspiration (ET), cloud cover (CC) (low, medium, high, and total), cloud base height (CBH), precipitable water vapor (PWV), and rainfall in the Beas River basin from 1980 to 2020. The findings indicate that the basin has been warming over the previous four decades. During the research period, the ET in the basin showed both increasing and declining patterns. The region's cloud cover conditions show a similar tendency. During the study period, particularly in April, the CBH is increasing in the region. During the months before the monsoon, the PWV is increasing in the area. The generation of hydropower and other operations that rely directly or indirectly on the availability of water may be impacted by rainfall that fluctuates over time. As a result, more thorough climate monitoring is necessary across the Beas River basin for better planning and management.

Keywords Air temperature · Evapotranspiration · Cloud cover · Cloud base height · Precipitable water vapor · Rainfall

3.1 Introduction

The hydrology of a basin is primarily affected by climate dynamics. For instance, rising air temperature increases the atmospheric water requirement, hence evaporation. The water cycle will likely intensify, precipitation patterns will change, glaciers and ice sheets will melt, the snow cover will shift, the mean sea level will rise, etc., and all of these changes might be made worse by more global warming (Schmidt and Nüsser 2012; Forsythe et al. 2014; Shrestha et al. 2015; Wijngaard et al. 2017; Schmidt and Nüsser 2017; Nie et al. 2018; Huss and Hock 2018; Ali

© The Author(s), under exclusive license to Springer Nature Switzerland AG 2023
S. Rani, *Climate, Land-Use Change and Hydrology of the Beas River Basin, Western Himalayas*, Advances in Asian Human-Environmental Research, https://doi.org/10.1007/978-3-031-29525-6_3

et al. 2019; Hassan et al. 2019; Krakauer et al. 2019; Nüsser et al. 2019; Hussain et al. 2020, 2021; Schmidt et al. 2020; Baig et al. 2021; Hussain et al. 2021; Nie et al. 2021; IPCC 2021; Liaqat et al. 2022). Additionally, it is anticipated that the precipitation patterns would alter both geographically and temporally. It is also notable that mountainous regions exhibit greater sensitivity to climate change as a result of their enormous water storage provided by snow and glaciers at high elevations as well as their location at the source of the majority of the world's large river networks (Xenarios et al. 2019). There is a need to analyze the trends in climate variables of the Beas River basin to assess their impacts on the basin hydrology. This chapter aims to discuss the trends and relationship of mean air temperature (T_{mean}), evapotranspiration (ET), cloud cover (CC) (low, middle, high, and total), cloud base height (CBH), precipitable water vapor (PWV), and rainfall of the Beas River basin during 1980–2020. Details of climate datasets and methods used in this chapter are given in the first chapter.

3.2 Trends in Air Temperature

3.2.1 Annual Trends

Air temperature is one of the key indicators of climate change. The Beas basin is located in the highly elevated topography of the Himalayas and source of water of livelihood, indicating the importance of assessing the changes in climate variables. An annual T_{mean} of 14.93 °C was found in the Beas basin during 1980–2020 (Table 3.1 and Fig. 3.1). Throughout the study, the highest annual T_{mean} of 16.26 °C was in 2016 and the lowest annual T_{mean} of about 13.84 °C in 1996 (Fig. 3.1). The study has found significant warming in annual T_{mean} (0.27 °C/decade) during the study period (Table 3.1 and Fig. 3.1).

3.2.2 Seasonal Trends

In the Beas basin, the observed T_{mean} is highest in monsoon, followed by pre-monsoon, during 1980–2020 (Table 3.1 and Fig. 3.2). High variation was found in the winter season T_{mean} (12.34%), followed by post-monsoon (6.02%), pre-monsoon (4.99%), and monsoon season (2.31%) (Table 3.1). These results are consistent with the study of Rani and Sreekesh (2018). T_{mean} showed a significant warming trend in monsoon (0.25 °C/decade) and post-monsoon (0.35 °C/decade) in the basin during the study period (Table 3.1).

Table 3.1 Descriptive statistics and trend of T_{mean} of the Beas basin

Time scale	Mean (°C)	Std. deviation (°C)	Coefficient of variation (%)	Trend (°C/decade)
Jan	4.24	1.04	24.59	0.04
Feb	6.95	1.34	19.29	4.30
Mar	11.20	1.42	12.67	4.70
Apr	16.44	1.45	8.82	5.60*
May	20.73	1.47	7.07	3.30
Jun	22.90	0.80	3.49	0.05
Jul	22.68	0.52	2.30	1.60
Aug	22.24	0.49	2.20	1.70*
Sep	20.15	0.73	3.61	2.60*
Oct	15.41	0.97	6.32	4.90***
Nov	10.22	0.82	8.04	3.50***
Dec	6.15	0.96	15.67	0.07
Winter	7.15	0.88	12.34	2.20
Pre-monsoon	20.12	1.00	4.99	0.36
Monsoon	21.73	0.50	2.31	0.25***
Post-monsoon	12.76	0.77	6.02	0.35***
Annual	14.93	0.57	3.79	0.27***

Sig at 0.10*, 0.05**, and 0.01***

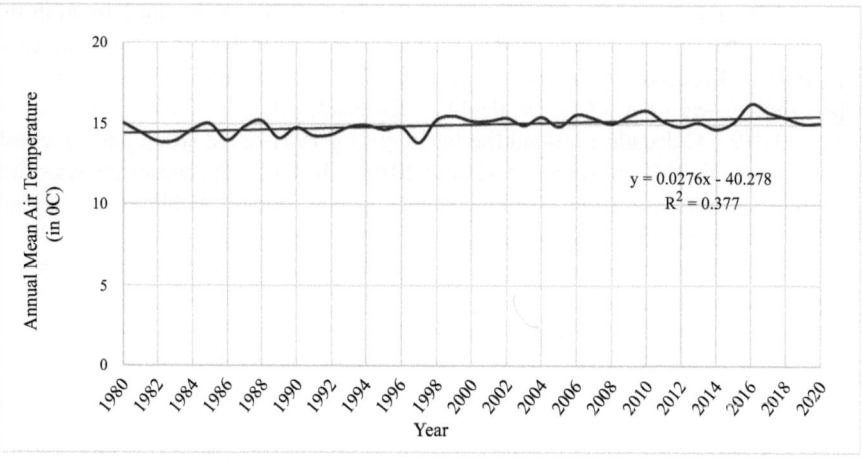

$$y = 0.0276x - 40.278$$
$$R^2 = 0.377$$

Fig. 3.1 Time series of annual T_{mean} of the Beas basin

3.2.3 Monthly Trends

Monthly T_{mean} of the basin varies between 4.24 °C (January) and 22.90 °C (June) during 1980–2020 (Table 3.1 and Fig. 3.3). The monthly trend analysis shows statistically significant warming in T_{mean} in April (0.56 °C/decade) and August–September (0.17 °C/decade) during the study period (Table 3.1). It may be attributed

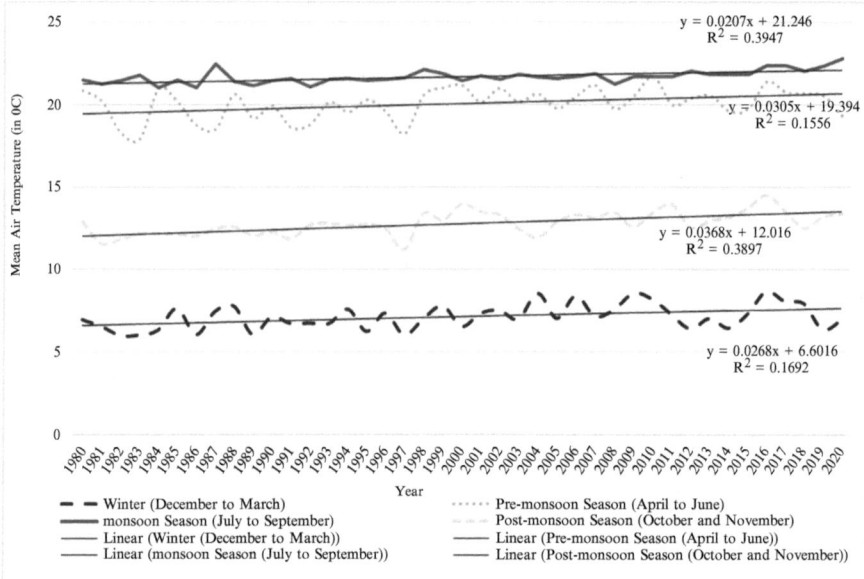

Fig. 3.2 Time series of seasonal T_{mean} of the Beas basin

to the rising mean T_{min} and T_{max} (Rani and Sreekesh 2018). A warming trend in the annual T_{mean} (0.061 °C/decade) with significant warming in T_{max} (0.1 °C/decade) and T_{min} (0.022 °C/decade) was observed during 1901–2019 (IMD 2019). Studies have also shown warming (T_{mean} = 0.044 ° C/decade; T_{max} = 0.051 ° C/decade; T_{min} = 0.019 ° C/decade) around the basin during 1941–2012 (Rani and Sreekesh 2018; Ray et al. 2019; Saxena and Mathur 2019). The lapse rate causes the warming rate to vary with elevation. Stronger warming was evident at higher elevations, probably as a result of elevation-dependent warming (EDW) (Liu et al. 2009; Krishnan et al. 2019). Highest warming in T_{min} is found compared to the T_{max} in the Western Indian Himalayas (WIH) (Dimri et al. 2018). Over the period of 1980–2020, warming during the winter has been at the fastest rate (up to 0.42 °C/decade) (Rani et al. 2022). Using extensive historical data spanning more than a century, Ren et al. (2017) discovered both an increase and decline in temperature trends throughout the Hindu-Kush Himalayas (HKH) region. The annual T_{mean} has shown a rising trend of about 0.2 °C/decade over the eastern side of the HKH range (Ren et al. 2017). Over northern India (the Sichuan Basin) and the Karakoram range during the summer, the warming rates in annual T_{mean} are less than 0.10 °C/decade (Forsythe et al. 2017). Although there are uncertainties in both observational studies and climate change modelling projections in the HKH, Gupta et al. (2020) observed the highest rate of warming at the highest altitude station (Kaza) (0.84 °C/decade) in the representative concentration pathway (RCP) at 8.5 over the Satluj River Basin (near the Beas River basin) (Mayewski et al. 2020). Hussain et al. (2021) showed a broad significant rising trend of 0.14 °C/decade for the maximum temperature for the period 1955–2016

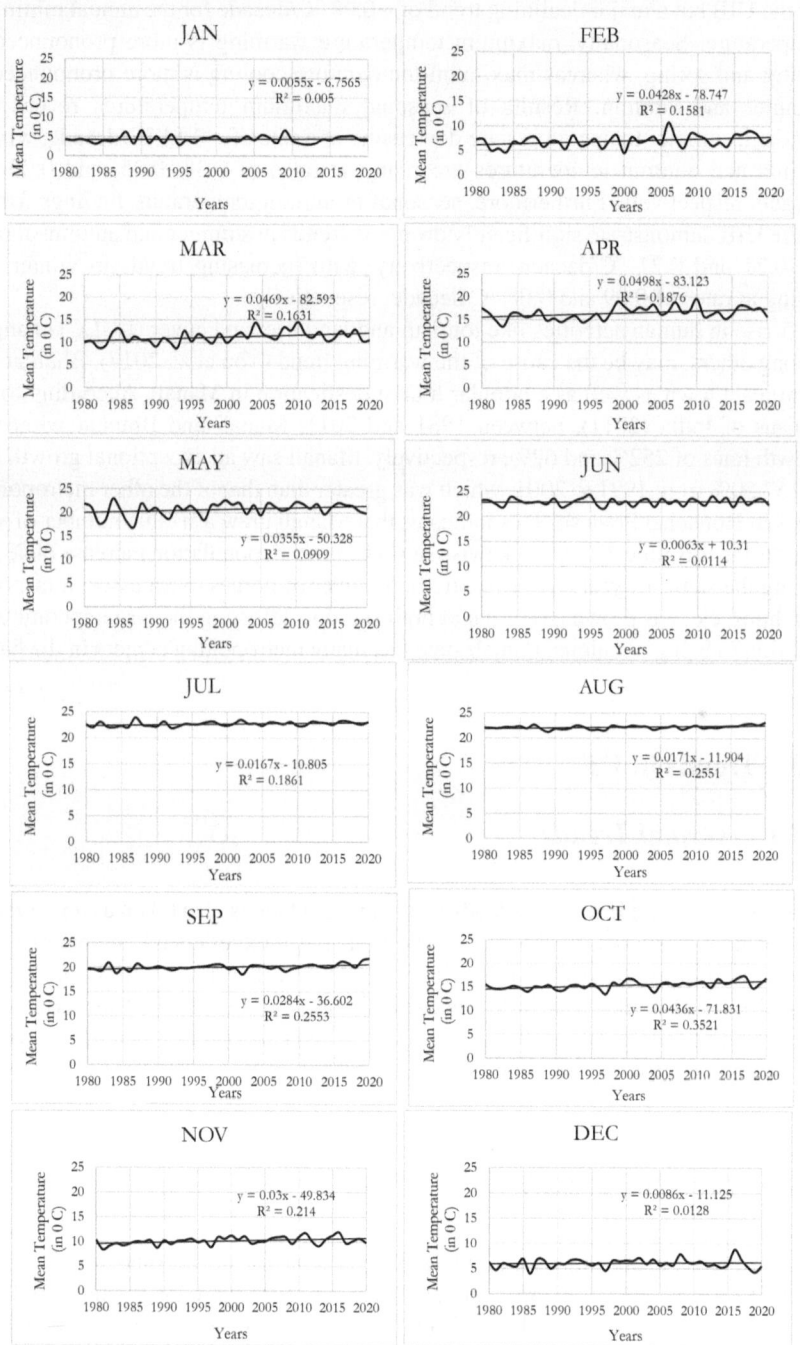

Fig. 3.3 Time series of monthly T_{mean} of the Beas basin

for the UIB but a major declining trend of −0.08 °C/decade for the annual minimum temperature. Seasonally, maximum temperature warming is more pronounced in winter and spring, whereas maximum temperature cooling is more pronounced in summer and autumn. Results of seasonal maximum temperatures reveal that although summer temperatures are decreasing at a rate of −0.14 °C/decade, winter, spring, and autumn temperatures are rising at rates of 0.38, 0.35, and 0.05 °C/decade, respectively. Furthermore, seasonal minimum temperature findings for the entire UIB demonstrate significantly dropping trends in summer and autumn at rates of -0.21 and 0.21 °C/decade, respectively, with increasing trends in winter and spring at rates of 0.09 and 0.08 °C/decade, respectively.

A rise in human activities like tourism and land use/land cover (LULC) changes, among others, may be the cause of the warming trend (You et al. 2017). Bhuntar is a significant town as well as a popular tourist destination in Manali. According to the Census of India (2011), between 1981 and 2011, Manali and Bhuntar witnessed growth rates of 252% and 62%, respectively. Manali saw an exceptional growth rate of 157.50% from 1991 to 2001, which was greater than that of the other metropolitan areas in Himachal Pradesh. This indicates that Manali drew a sizable number of visitors during this time. Bhuntar likewise saw a significant population increase of 43.3% during the same 10 years. As a result, the number of homes, businesses, transportation hubs, etc. has grown over time at both stations, affecting the air temperature. In addition to being populous, Punjab state has many metropolitan centers in the basin.

3.3   Trends in ET

3.3.1   Annual Trends

The warming trend in any area leads to rising ET which is also taken as an indicator of climate change around the globe. The Beas basin experienced an annual ET of 28.68 mm during the period 1980–2020 (Table 3.2). During the study period, the annual ET had a relatively little standard deviation and coefficient of variation of 0.91 mm and 3.16%, respectively (Table 3.2). Over the studied period, the yearly ET exhibits no discernible trend (Table 3.2 and Fig. 3.4).

3.3.2   Seasonal Trends

Winter, pre-monsoon, monsoon, and post-monsoon ET in the Beas basin were 6, 9.05, 9.73, and 2.44 mm, respectively, between 1980 and 2020 (Table 3.2 and Fig. 3.5). High variation in ET was found in the post-monsoon (20.6%), followed by pre-monsoon (7.80%), monsoon (4.6%), and winter (3.7%) (Table 3.2). ET showed a significant declining trend in pre-monsoon (−0.04 °C/decade) in the study area during the study period (Table 3.2).

Table 3.2 Descriptive statistics and trend of ET of the Beas basin

Time scale	Mean (mm)	Std. deviation (mm)	Coefficient of variation (%)	Trend (mm/decade)
Jan	1.09	0.08	7.76	0.00
Feb	1.60	0.11	6.99	0.02
Mar	2.36	0.15	6.51	0.02
Apr	2.83	0.26	9.29	−0.07*
May	2.98	0.23	7.84	−0.07
Jun	3.26	0.22	6.87	−0.03
Jul	3.33	0.24	7.33	0.02
Aug	3.25	0.30	9.12	−0.06
Sep	3.20	0.13	4.18	−0.05*
Oct	2.31	0.15	6.53	0.02
Nov	1.43	0.10	6.71	0.01
Dec	1.03	0.07	6.34	0.03***
Winter	6.00	0.22	3.73	0.04
Pre-monsoon	9.05	0.71	7.80	−0.21*
Monsoon	9.73	0.45	4.61	−0.04
Post-monsoon	2.44	0.50	20.60	0.11
Annual	28.68	0.91	3.16	−0.15

Sig at 0.10*, 0.05**, and 0.01***

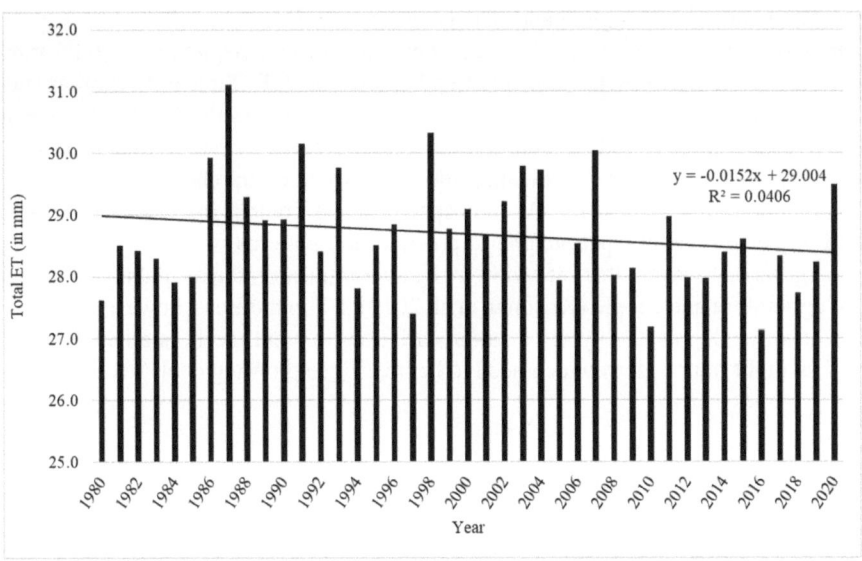

Fig. 3.4 Time series of annual ET of the Beas basin

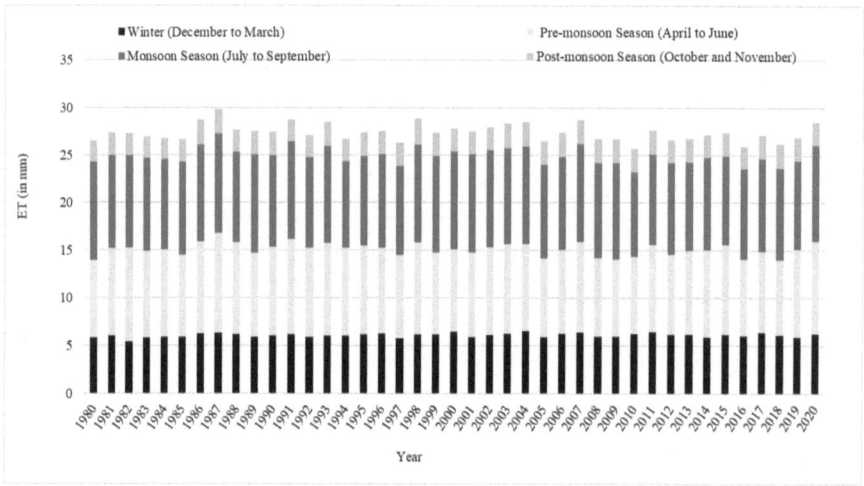

Fig. 3.5 Time series of seasonal ET of the Beas basin

3.3.3 Monthly Trends

Additionally, the basin's monthly ET varied from 4.24 mm in January to 22.9 mm in June from 1980 to 2020 (Table 3.2 and Fig. 3.6). From June to September, the coefficient of variation reveals very little fluctuation in ET (Table 3.2). In the study period, a statistically significant increasing trend in December (0.03 mm/decade) and a declining trend in April (−0.07 mm/decade) and September (−0.05 mm/decade) are revealed by the seasonal trend analysis of ET (Table 3.2). Similar substantial trends of declining mean ET were also seen across the Indian Himalayas (IH) (mean rate of −0.002 mm/decade, ranging between −0.12 and 0.06 mm/decade) (Rani et al. 2022), indicating the necessity to comprehend the connection between temperature and ET. As atmospheric water content rises due to warming and higher ET, future spatiotemporal precipitation patterns will shift. Particularly during the dry season, this may have a negative impact on the availability of water for agriculture and people (Krishnan et al. 2019). ET in the Beas, however, is trending downward, presumably as a result of variations in wind speed, relative humidity, and sunlight hours, all of which need to be investigated (Wang et al. 2012).

3.4 Trends in Cloud Cover

3.4.1 Annual Trends

Cloud cover conditions are an indicator of climate change in an area which is affected by both temperature and ET. The Beas basin experienced an annual low cloud cover (LCC) of 14.90% during the period 1980–2020, while the middle

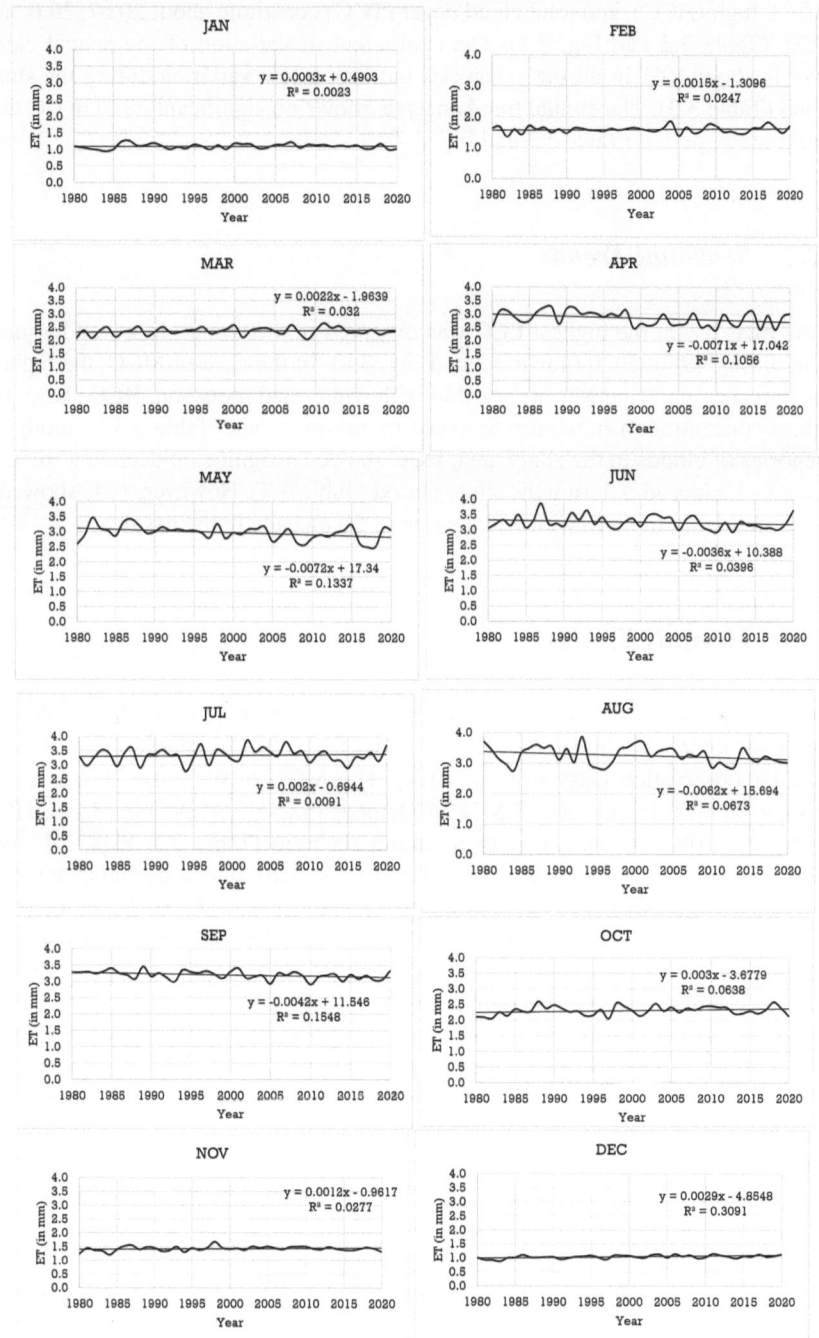

Fig. 3.6 Time series of monthly ET of the Beas basin

(MCC), high (HCC), and total cloud cover (TCC) constitute about 20.07, 26.6, and 0.27% (Table 3.3 and Fig. 3.7). The coefficient of variation of the annual cloud cover is about 10% in all the categories indicating low variation during the study period (Table 3.3). The annual trend analysis shows no significant trend in CC during the study period (Table 3.3 and Fig. 3.7).

3.4.2   Seasonal Trends

In the Beas basin, the highest LCC was observed in the winter, followed by monsoon, during 1980–2020 (Table 3.3 and Fig. 3.8). In the case of MCC, the highest and almost equal value was observed in both winter and monsoon. HCC shows the highest concentration in winter, followed by pre-monsoon (Table 3.3). Among all categories of clouds in the study area, HCC showed a significant declining trend in winter (−1%/decade) during the study period (Table 3.3). However, TCC showed a significant rising trend in winter at the rate of 5%/decade in the basin.

3.4.3   Monthly Trends

Between 1980 and 2020, the basin's monthly LCC percentage varied from 4.90% in November to 36.44% in August (Table 3.3 and Fig. 3.9). Another research showed a similar observation (Jaswal et al. 2017). The MCC of the basin shows values between 30.66% in July and 7.51% in October (Table 3.3 and Fig. 3.10). HCC above the basin also demonstrates a similar scenario (Table 3.3, Figs. 3.11 and 3.12). A statistically significant rising trend in January and a declining trend in November can be shown in the monthly trend analysis of LCC. HCC, on the other hand, exhibits a growing tendency in November over the study period (Table 3.3). During the study period, the TCC shows a growing tendency overall in March, November, and a declining trend in December (Table 3.3).

   TCC above the Indian Himalayas has been significantly declining since 1980 at a mean rate of 0.0004%/decade (Rani et al. 2022). The overall amount of clouds in the area have significantly decreased throughout the whole winter (Dec-Mar) (Rani et al. 2022). The monsoon, which decreased at a rate of 1.22% per decade between 1961 and 2010, is the major cause of the annual mean LCC considerable decreasing tendency, which is discovered at a rate of 0.45%/decade (Jaswal et al. 2017). Data analysis reveals a substantial negative link between the highest temperature and the range of temperatures throughout the day in India from 1961 to 2010, and a large positive correlation between the number of rainy days (Jaswal et al. 2017). A detailed understanding of the cause-and-effect relationship between rainfall and cloud cover across the basin is required using fine-resolution meteorological data.

Table 3.3 Descriptive statistics and trend of cloud cover of the Beas basin

Time scale	Mean (%)	Std. deviation (%)	Coefficient of variation (%)	Trend (%/decade)
LCC				
Jan	15.02	4.94	32.90	1.00*
Feb	13.95	5.17	37.08	0.05
Mar	9.80	3.62	36.95	−1.00*
Apr	6.85	1.91	27.81	0.00
May	6.95	1.66	23.86	0.00
Jun	15.17	5.65	37.26	0.00
Jul	35.54	6.63	18.64	−1.00
Aug	36.44	6.01	16.50	−1.00
Sep	19.56	4.96	25.36	0.00
Oct	6.59	2.78	42.11	0.00
Nov	4.90	2.78	56.78	1.00*
Dec	9.10	4.57	50.23	1.00
Winter	11.98	2.49	20.74	0.00
Pre-monsoon	9.68	2.01	20.71	0.00
Monsoon	30.41	3.76	12.37	0.00
Post-monsoon	5.66	2.02	35.69	0.00
Annual	14.90	1.55	10.38	0.00
MCC				
Jan	27.59	8.71	31.57	0.00
Feb	29.88	9.11	30.47	0.00
Mar	24.95	6.96	27.88	−2.00
Apr	17.49	4.69	26.82	−1.00
May	13.56	3.33	24.52	0.00
Jun	16.20	4.24	26.20	1.00
Jul	30.66	7.22	23.54	0.00
Aug	28.76	6.31	21.93	1.00
Sep	15.66	4.37	27.89	0.00
Oct	7.51	3.89	51.76	0.00
Nov	10.51	5.73	54.54	1.00
Dec	17.83	7.69	43.11	−2.00
Winter	25.07	4.46	17.80	−1.00
Pre-monsoon	15.76	1.93	12.27	0.00
Monsoon	25.02	3.83	15.31	1.00
Post-monsoon	9.02	3.71	41.14	1.00
Annual	20.07	2.18	10.88	0.00
HCC				
Jan	26.10	8.50	32.57	0.00
Feb	31.02	11.93	38.47	1.00
Mar	37.22	8.82	23.70	−2.00
Apr	35.54	7.90	22.23	0.00
May	29.05	9.17	31.56	−2.00

(continued)

Table 3.3 (continued)

Time scale	Mean (%)	Std. deviation (%)	Coefficient of variation (%)	Trend (%/decade)
Jun	20.00	7.30	36.52	1.00
Jul	39.88	13.00	32.59	−1.00
Aug	39.68	14.81	37.33	3.00
Sep	14.46	6.89	47.64	1.00
Oct	9.46	7.77	82.16	0.00
Nov	15.27	7.65	50.09	2.00*
Dec	22.12	9.04	40.88	−3.00
Winter	29.12	4.16	14.29	−1.00*
Pre-monsoon	28.17	4.24	15.04	0.00
Monsoon	31.32	7.76	24.78	1.00
Post-monsoon	12.39	6.23	50.27	1.00
Annual	26.61	2.85	10.73	0.00
TCC				
Jan	0.26	0.09	32.57	0.00
Feb	0.31	0.12	38.47	1.00
Mar	0.37	0.09	23.70	−2.00*
Apr	0.36	0.08	22.23	0.00
May	0.29	0.09	31.56	−2.00
Jun	0.20	0.07	36.52	1.00
Jul	0.40	0.13	32.59	−1.00
Aug	0.40	0.15	37.33	3.00
Sep	0.14	0.07	47.64	1.00
Oct	0.09	0.08	82.16	0.00
Nov	0.15	0.08	50.09	2.00*
Dec	0.22	0.09	40.88	−3.00*
Winter	0.29	0.04	14.29	5.10*
Pre-monsoon	0.28	0.04	15.04	0.00
Monsoon	0.31	0.08	24.78	1.00
Post-monsoon	0.12	0.06	50.27	1.00
Annual	0.27	0.03	10.73	0.00

Sig at 0.10*, 0.05**, and 0.01***

3.5   Trends in Cloud Base Height

3.5.1   Annual Trends

Between 1980 and 2020, the Beas basin had an average CBH of 2612 m above the surface (Table 3.4). During the study period, the yearly CBH's standard deviation and coefficient of variation were 192 m and 7%, respectively (Table 3.4), which indicates a relatively slight change. The annual CBH shows no significant trend during the study period (Table 3.4 and Fig. 3.13).

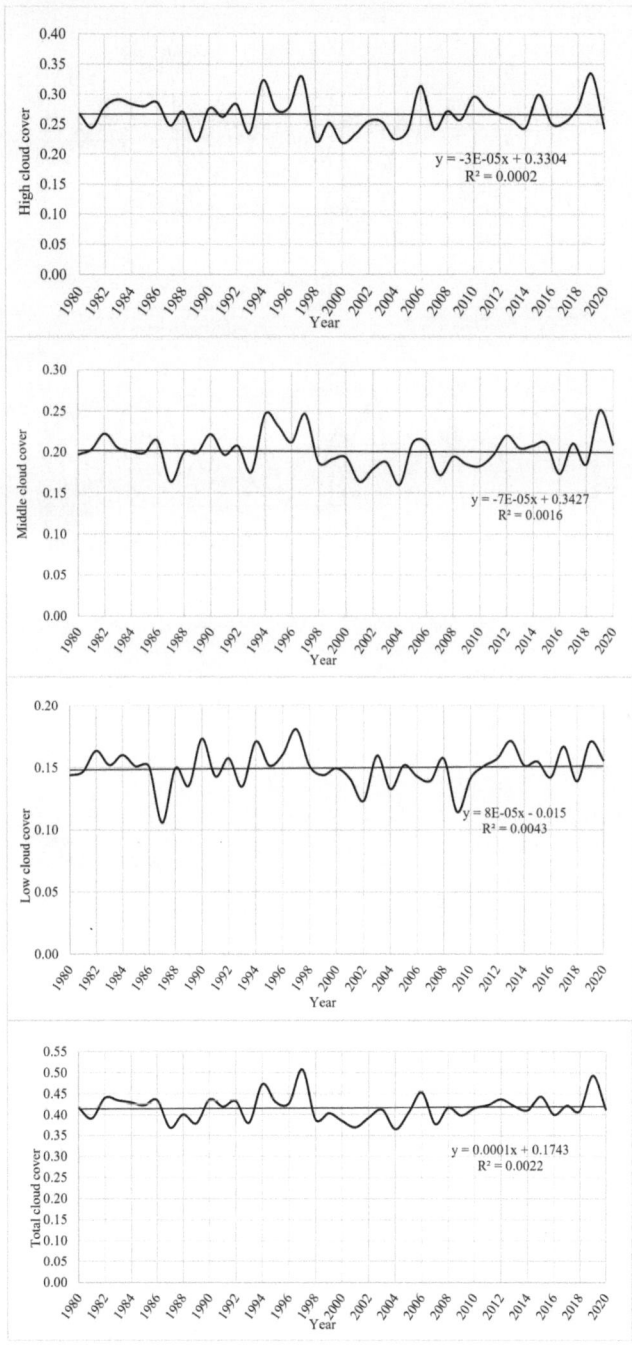

Fig. 3.7 Time series of annual low, middle, high, and total cloud cover of the Beas basin

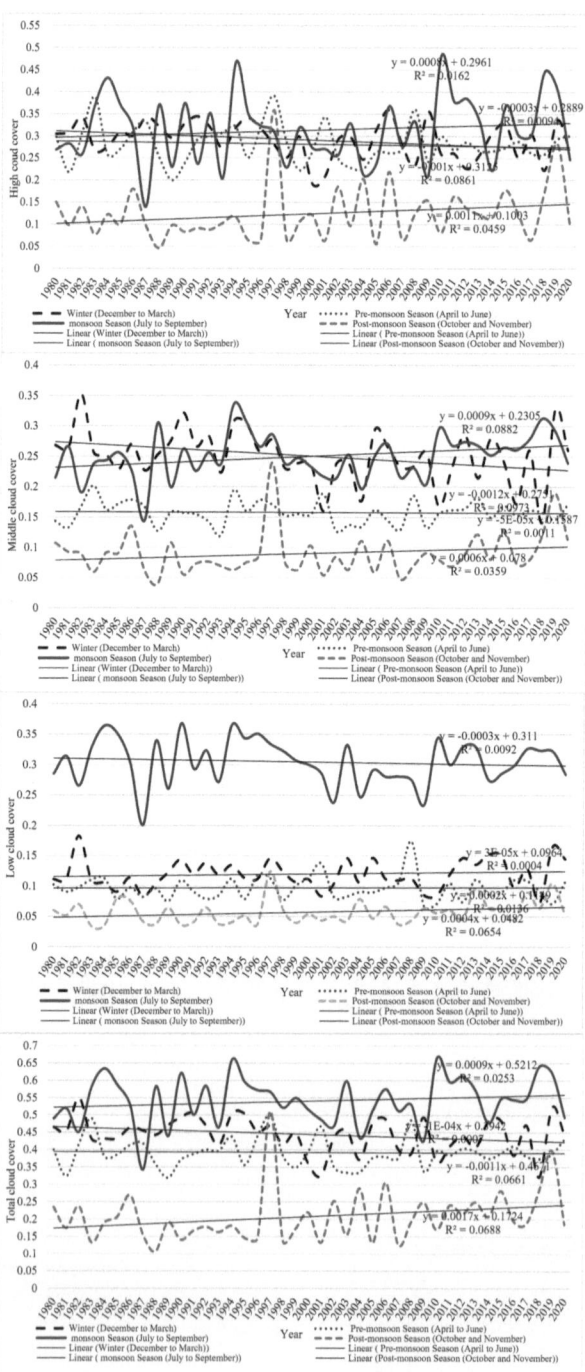

Fig. 3.8 Time series of seasonal low, middle, high, and total cloud cover of the Beas basin

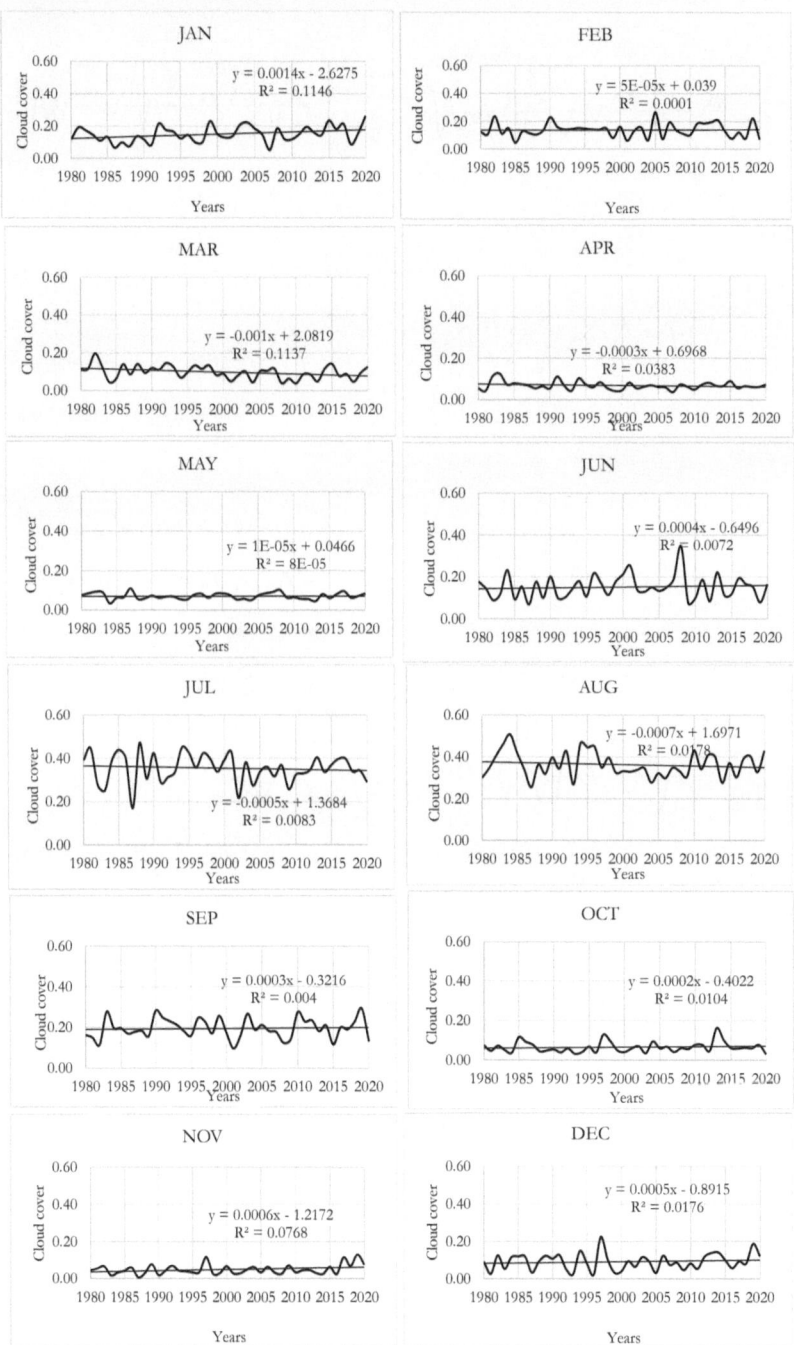

Fig. 3.9 Time series of monthly low cloud cover of the Beas basin

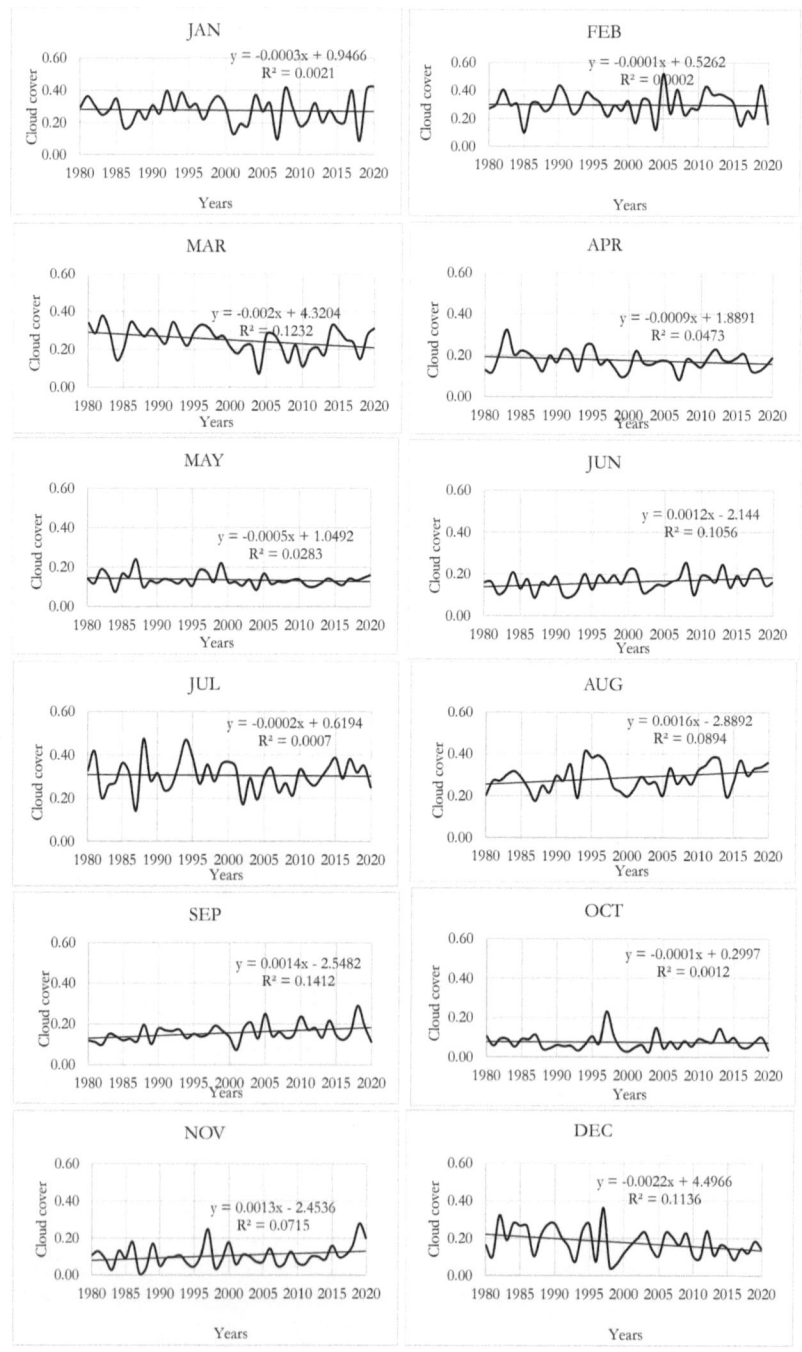

Fig. 3.10 Time series of monthly middle cloud cover of the Beas basin

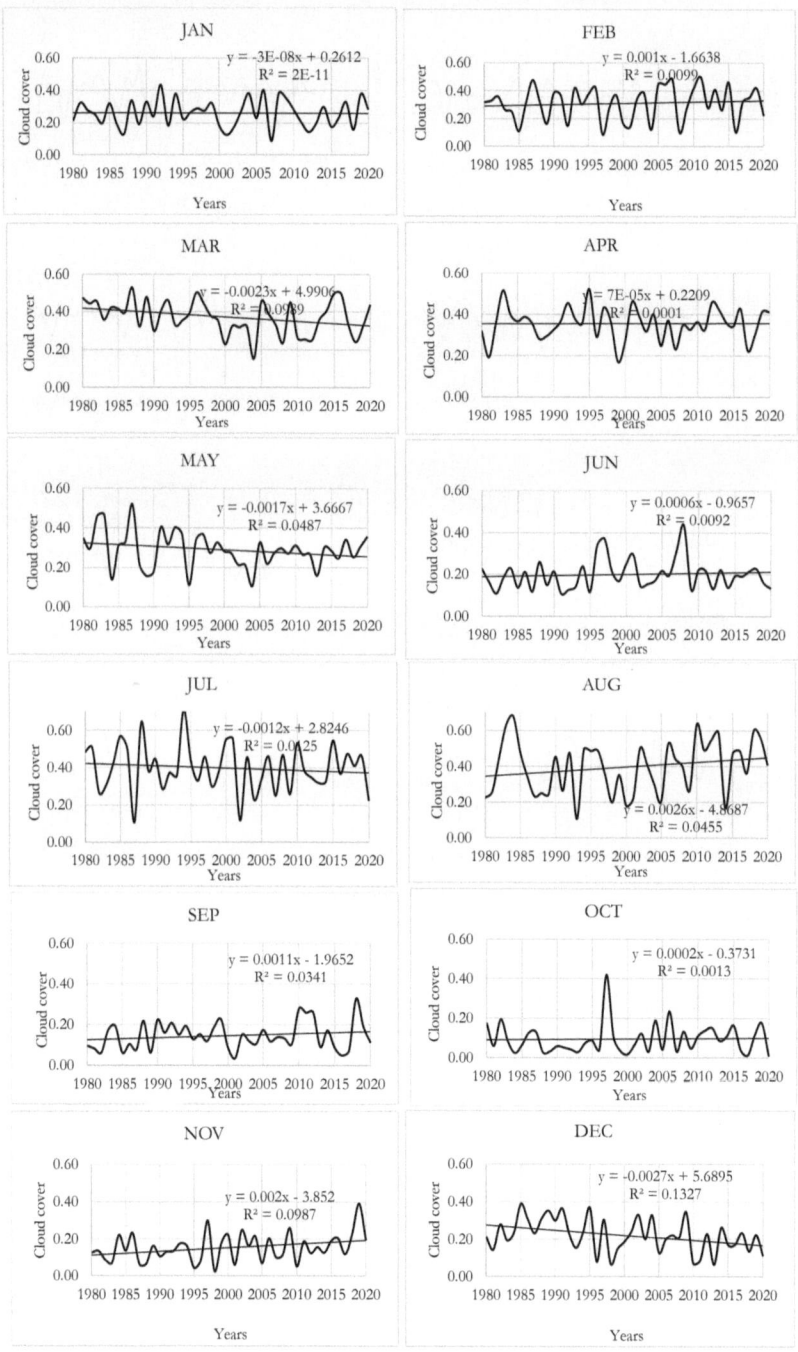

Fig. 3.11 Time series of monthly high cloud cover of the Beas basin

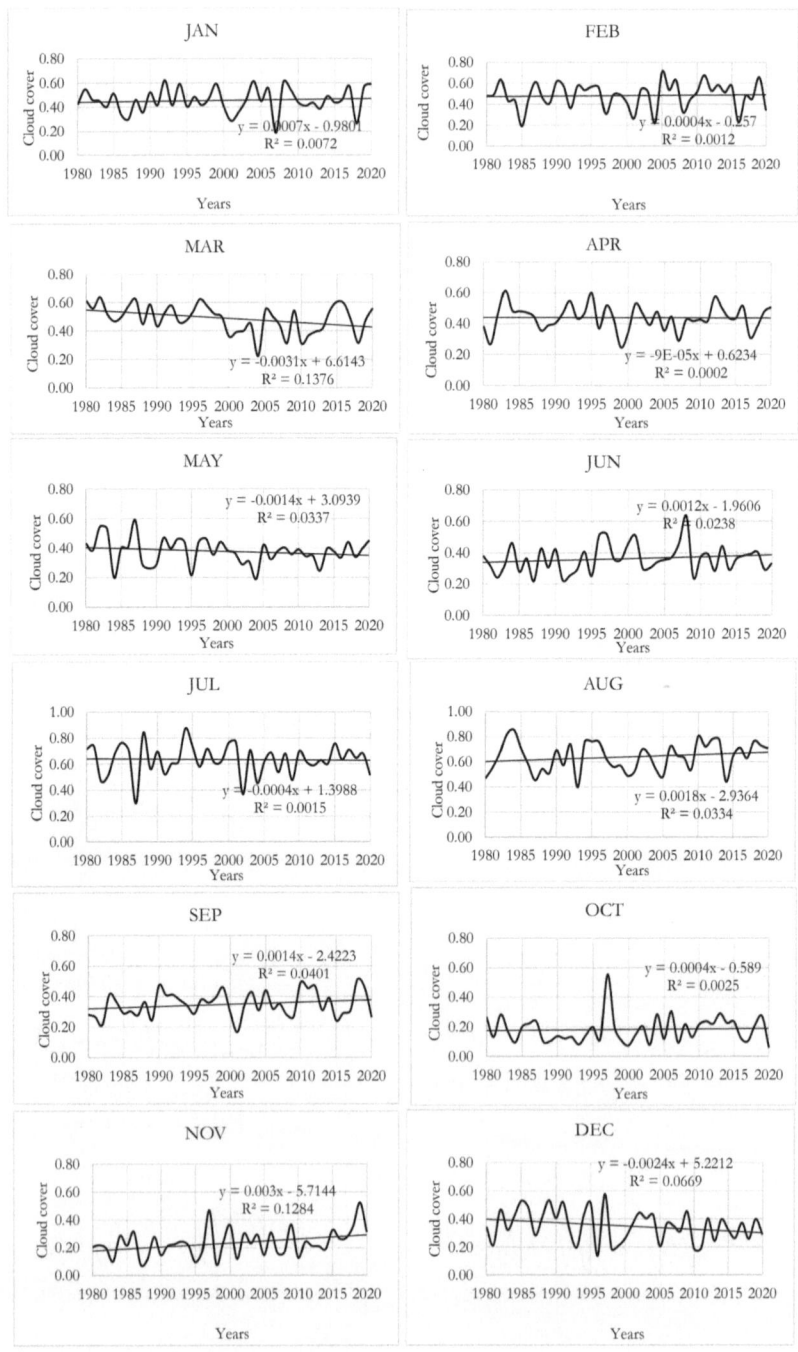

Fig. 3.12 Time series of monthly total cloud cover of the Beas basin

Table 3.4 Descriptive statistics and trend of cloud base height of the Beas basin

Time scale	Mean (m)	Std. deviation (m)	Coefficient of variation (%)	Trend (m/decade)
Jan	2768	548	20	−94
Feb	2763	562	20	63
Mar	3337	597	18	103
Apr	3774	471	12	104*
May	3663	346	9	10
Jun	2234	554	25	−78
Jul	945	252	27	−53
Aug	847	175	21	11
Sep	1380	316	23	−49
Oct	2512	656	26	6
Nov	3549	843	24	173
Dec	3569	759	21	−97
Winter	3109	326	10	−6
Pre-monsoon	3223	272	8	13
Monsoon	1057	147	14	−30
Post-monsoon	3031	519	17	89
Annual	2612	192	7	8

Sig at 0.10*, 0.05**, and 0.01***

3.5.2 Seasonal Trends

Pre-monsoon has the greatest CBH in the Beas basin from 1980 to 2020, followed by winter, post-monsoon, and monsoon (Table 3.4 and Fig. 3.13). Following the monsoon and moving into the winter, post-monsoon saw the most fluctuation in CBH (Table 3.4). The findings revealed insignificant increase in seasonal CBH over the study period in the study area (Table 3.4 and Fig. 3.14).

3.5.3 Monthly Trends

Between 1980 and 2020, the basin's monthly CBH varied from 847 m in August to 3569 m in December (Table 3.4 and Fig. 3.15). From March to May throughout the study period, the coefficient of variation of CBH shows comparatively little fluctuation (Table 3.4). The monthly trend analysis reveals a statistically significant rising trend in CBH of April at a rate of 104 m/decade over the studied period.

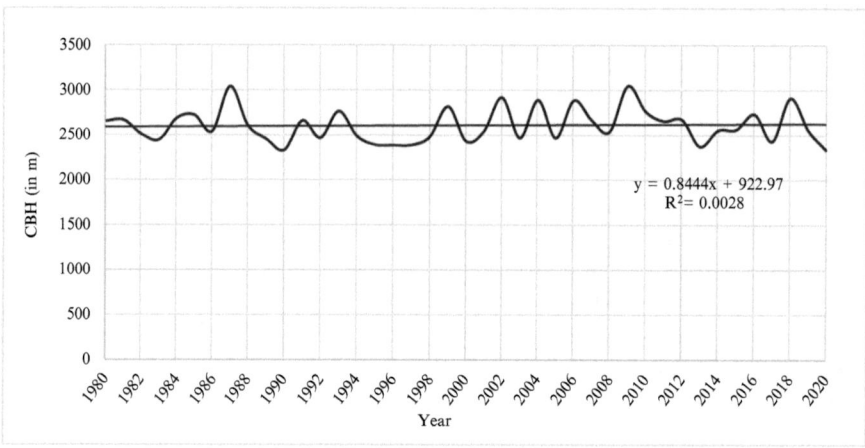

Fig. 3.13 Time series of annual cloud base height of the Beas basin

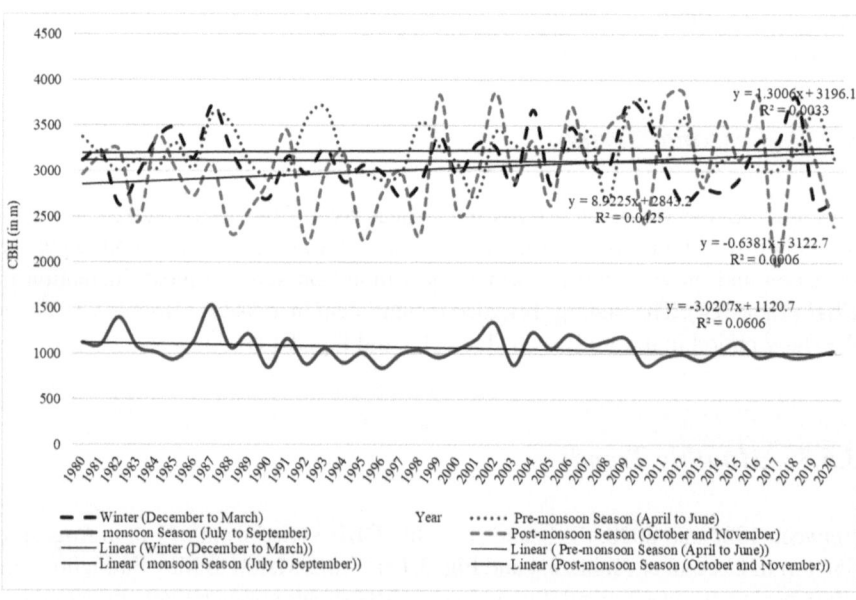

Fig. 3.14 Time series of seasonal cloud base height of the Beas basin

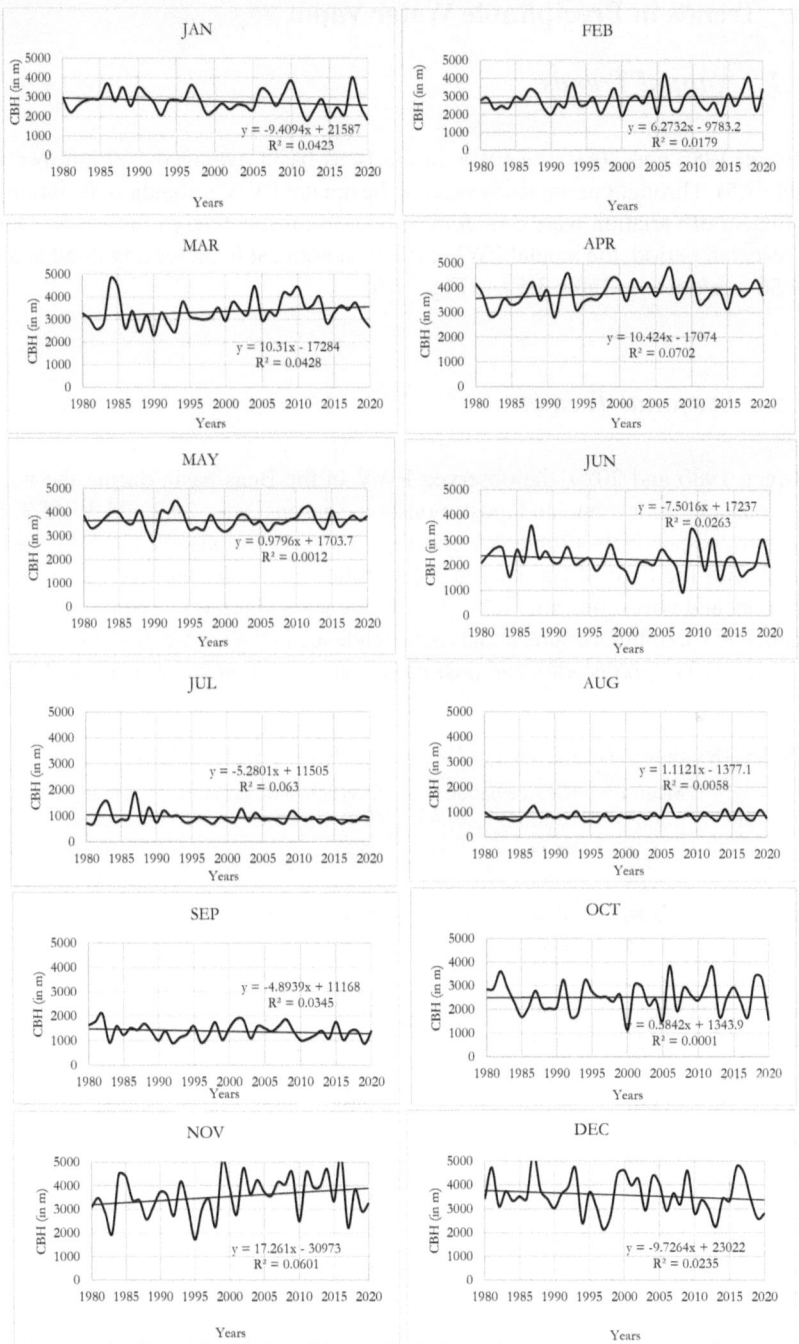

Fig. 3.15 Time series of monthly cloud base height of the Beas basin

3.6   Trends in Precipitable Water Vapor

3.6.1   Annual Trends

Between 1980 and 2020, the PWV in the Beas basin averaged 100 mm per year (Table 3.5). Throughout the study period, the annual PWV's standard deviation and coefficient of variation were 3.66 mm and 3.64%, respectively (Table 3.5). During the research period, the annual PWV exhibits a noticeable increasing trend at a rate of 1.56 mm/decade (Table 3.5 and Fig. 3.16).

3.6.2   Seasonal Trends

Between 1980 and 2020, the observed PWV in the Beas basin during the winter, pre-monsoon, monsoon, and post-monsoon seasons was 20.2, 23.3, 43.3, and 10.6 mm, respectively (Table 3.5 and Fig. 3.17). Pre-monsoon (10.5%) was the season with the highest variance in PWV, followed by the monsoon season, pre-monsoon, and post-monsoon (Table 3.5). In the studied region throughout the study period, PWV exhibited a substantial rising tendency in the winter (0.42 mm/decade), monsoon (0.54 mm/decade), and post-monsoon (0.22 mm/decade) (Table 3.5).

Table 3.5 Descriptive statistics and trend of precipitable water vapor of the Beas basin

Time scale	Mean (mm)	Std. deviation (mm)	Coefficient of variation (%)	Trend (mm/decade)
Jan	4.55	0.45	9.83	0.06
Feb	5.05	0.51	10.08	0.18
Mar	5.86	0.58	9.90	0.06
Apr	6.08	0.86	14.18	0.01
May	6.74	1.19	17.64	−0.11
Jun	10.56	1.76	16.68	0.18
Jul	15.12	0.83	5.51	0.14
Aug	15.05	0.37	2.47	0.08
Sep	13.07	0.92	7.03	0.32*
Oct	7.77	1.12	14.47	0.39*
Nov	5.89	0.61	10.42	0.18*
Dec	4.77	0.47	9.92	0.06
Winter	20.27	1.29	6.34	0.42*
Pre-monsoon	23.39	2.45	10.47	0.04
Monsoon	43.32	1.51	3.48	0.54**
Post-monsoon	10.63	0.97	9.11	0.22*
Annual	100.50	3.66	3.64	1.56***

Sig at 0.10*, 0.05**, and 0.01***

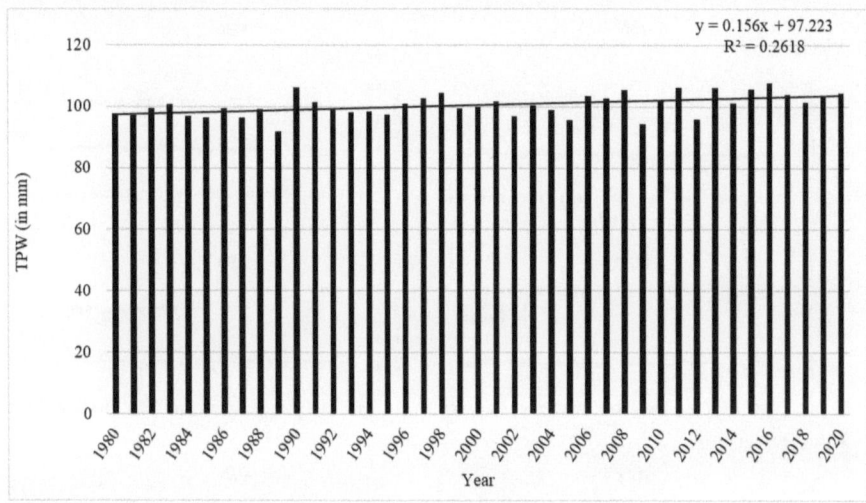

Fig. 3.16 Time series of annual precipitable water vapor of the Beas basin

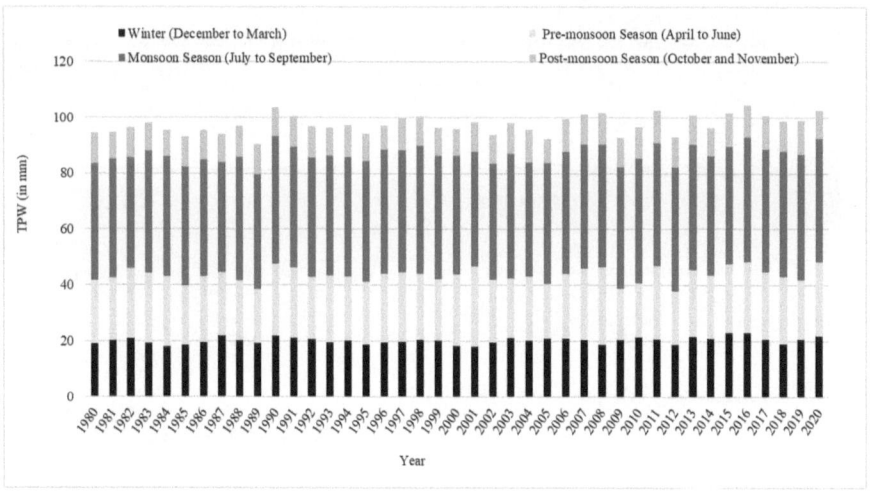

Fig. 3.17 Time series of seasonal precipitable water vapor of the Beas basin

3.6.3   Monthly Trends

Between 1980 and 2020, the basin's monthly PWV varied from 4.55 mm in January to 15.12 mm in July (Table 3.5 and Fig. 3.18). The monthly PWV fluctuation from July to August is quite small, according to the coefficient of variation (Table 3.5). During the study period, the monthly trend analysis of PWV reveals a statistically significant rising trend in September (0.32 mm/decade), October (0.39 mm/decade),

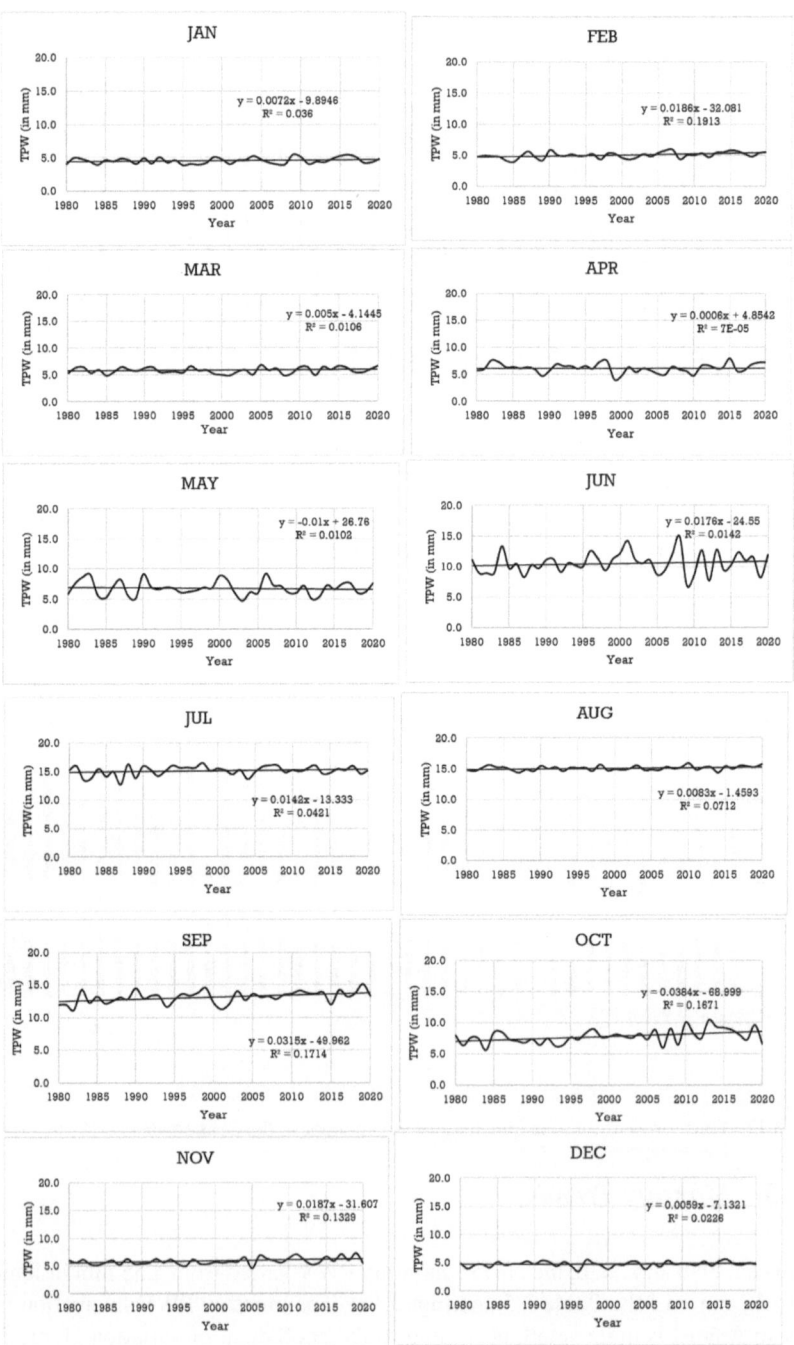

Fig. 3.18 Time series of monthly precipitable water vapor of the Beas basin

and November (0.18 mm/decade) (Table 3.5). Rising PWV is possibly due to rising ET in the area. However, ET is not showing a significant rise in the area. There is a need to explore it in detail with fine datasets. Rising PWV can be an indicator of increasing precipitation in the study area. Between 1979 and 2001, the magnitude of the global PWV increase was found to be about 0.26 mm/decade (Bengtsson et al. 2004). Global trends of total column PWV increased by 2.8 ± 0.8% during 1996–2002 (Wagner et al. 2006). In the period 1871–2010, the global mean water vapor increased by 0.0864 ± 0.01 kg m²/decade (Zhang et al. 2013). Consequently, PWV also provides strong positive feedback for global warming as well as other greenhouse gases (carbon dioxide, methane, and ozone) (Held and Soden 2000; Zhang et al. 2013).

3.7   Trends in Rainfall

3.7.1   Annual Trends

Between 1980 and 2020, the basin saw a mean annual rainfall of 46.77 mm (Table 3.6). The region's maximum and minimum annual rainfall were recorded as 71.9 mm in 1988 and 34 mm in 1987, respectively (Fig. 3.19). There was less variance in the annual rainfall throughout the analysis period, as indicated by the annual rainfall's standard deviation and coefficient of variation is 8.27 mm and 17.68%, respectively (Table 3.6). The annual trend analysis for the study area revealed no change in the annual rainfall between 1980 and 2020.

3.7.2   Seasonal Trends

Between 1980 and 2020, the observed mean rainfall for the winter, pre-monsoon, monsoon, and post-monsoon seasons was 12.4, 7.8, 25, and 2.3 mm, respectively (Fig. 3.20). According to the coefficient of variation of rainfall, the post-monsoon season has the most variance (68.8%), followed by the pre-monsoon (30.6%), the winter (27.9%), and the monsoon season (26.6%) (Table 3.6). Rani and Sreekesh (2018) report similar results. There is no discernible pattern in the region according to the seasonal trend study of seasonal rainfall (Table 3.6).

3.7.3   Monthly Trends

In the basin, the mean monthly rainfall varied from 0.8 mm in November to 11.4 mm in July between 1980 and 2020 (Table 3.6). It demonstrates that the monsoon season's month of July saw the highest rainfall. The monthly rainfall coefficient of

Table 3.6 Descriptive statistics and trend of rainfall of the Beas basin

Time scale	Mean (mm)	Std. deviation (mm)	Coefficient of variation (%)	Trend (mm/decade)
Jan	2.87	1.59	55.31	0.21
Feb	4.23	2.01	47.47	0.09
Mar	3.71	1.98	53.47	−0.53*
Apr	2.40	1.29	53.66	−0.30*
May	1.74	0.98	56.22	−0.22*
Jun	3.63	1.97	54.41	0.13
Jul	11.42	4.17	36.54	−1.11*
Aug	9.71	2.90	29.90	0.17
Sep	3.84	2.51	65.30	0.22
Oct	0.83	0.82	98.91	−0.06
Nov	0.85	0.89	105.46	0.06
Dec	1.53	1.22	79.49	−0.23
Winter	12.37	3.45	27.88	−0.48
Pre-monsoon	7.80	2.39	30.62	−0.38
Monsoon	25.00	6.57	26.29	−0.71
Post-monsoon	2.34	1.61	68.61	−0.19
Annual	46.77	8.27	17.68	-1.58

Sig at 0.10*, 0.05**, and 0.01***

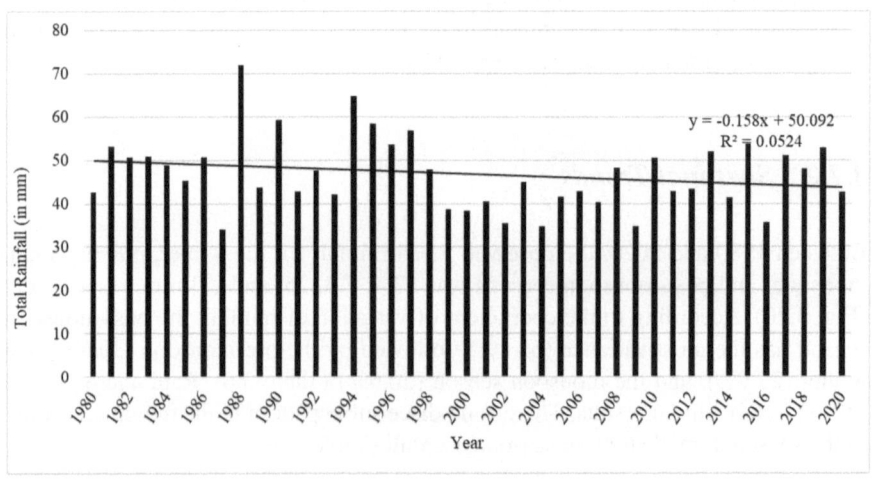

Fig. 3.19 Time series of annual rainfall of the Beas basin

variation indicates that there is a comparatively high variance in November (105%) and a low variation in August (30%) (Table 3.6). The examination of monthly trends reveals a considerable downward pattern in rainfall of March, April, May, and July (Table 3.6). March has a comparatively high rate of change in rainfall in the basin. If this magnitude of a downward trend sustains, it might eventually have a

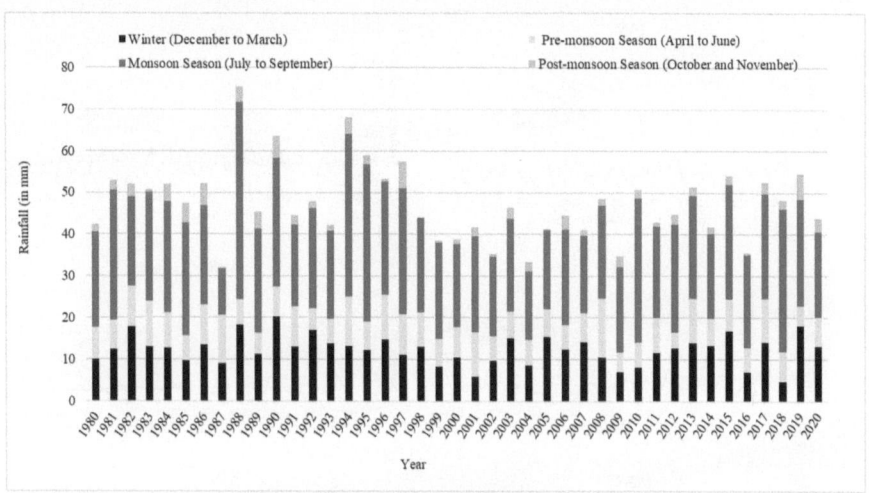

Fig. 3.20 Time series of seasonal rainfall of the Beas basin

significant impact on the Beas River's hydrology. The findings are in contrast to Kumar et al. (2010), who reported an increase in rainfall at the national level in June, July, and September and a declining pattern in August.

Over India, no consistently changing rainfall has been noted (Babar and Ramesh 2013; Kumar et al. 2018). Rainfall in the IHR has significantly decreased (−0.32 mm/decade) over the previous four decades (−1.13 to 0.14 mm/decade) (Rani et al. 2022). Over the IHR, the winter months had the greatest reduction in rainfall (up to −1.10 mm/decade) (Rani et al. 2022). In years 1957–2007, Singh and Mal (2014) noted a decreasing trend in the state of Uttarakhand's annual rainfall at high elevations (located in the WIH). While the winter rainfall exhibits diverse tendencies, the monsoon intensified in low elevations. In the Himalayas, summer precipitation has been trending downward (Schickhoff et al. 2016). In the Indus basin, Krakauer et al. (2019) found a 15% increase in mean precipitation between 1891 and 2016. For the more recent period, going back to 1958 or 1979, no clear precipitation trend was observed. Hussain et al. (2021) estimate that annual precipitation increases at a rate of 2.74 mm/decade, compared to rates of 1.18, 2.06, and 0.62 mm/decade for the winter, summer, and autumn of the UIB, respectively. Liaqat et al. (2022) found that between 1995 and 2017, the annual precipitation in the UIB increased only modestly, while the amount of precipitation during the winter months increased significantly. Changes in rainfall may have a direct impact on the hydrology of the Beas River basin, which would have an impact on the livelihood of a larger population. Therefore, frequent monitoring of the rainfall in the basin would be helpful for effective planning in the region to maintain the population's ability to sustain their way of life (Fig. 3.21).

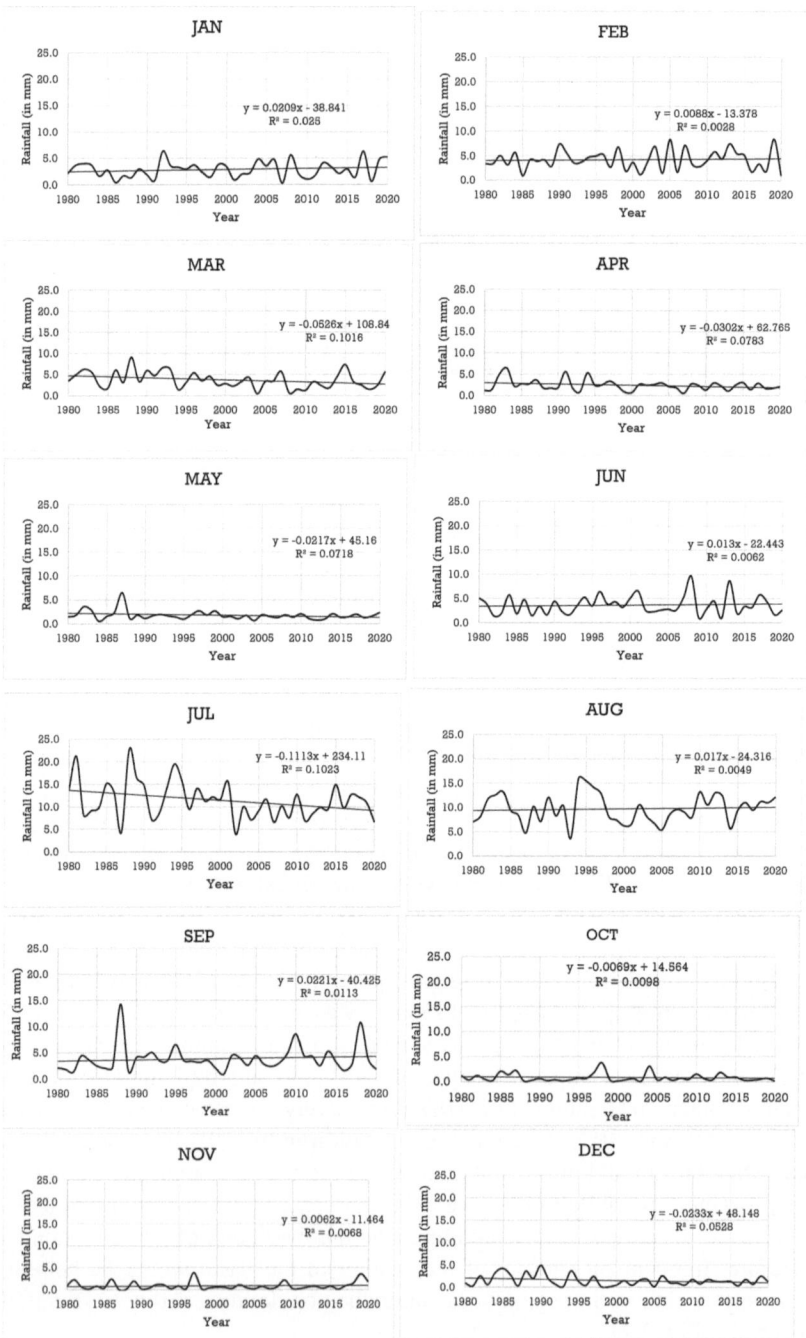

Fig. 3.21 Time series of monthly rainfall of the Beas basin

3.8   Trends in Percent Share of Total Rainfall

3.8.1   Seasonal Trends

To comprehend variations in the percent share of seasonal variability in rainfall in the basin, the trend in seasonal precipitation is evaluated. The observed mean rainfall percentage share for the winter, pre-monsoon, monsoon, and post-monsoon seasons, respectively, was 3%, 26%, 53%, and 5% between 1980 and 2020 (Fig. 3.22), showing that the monsoon season had the highest share, followed by the pre-monsoon season. The coefficient of variation of rainfall shows that the post-monsoon (61%) season has the most variability, followed by the pre-monsoon (30.3%), the winter (23%) season, and the monsoon season (14%) (Table 3.7).

3.8.2   Monthly Trends

In the period 1980–2020, the mean monthly rainfall varied from 1.8% in November to 24.6% in July (Table 3.7 and Fig. 3.23). It shows how much of the annual rainfall was contributed by July. The monthly rainfall's coefficient of variation reveals a comparatively low variance in August (30%) and a very large variation in November (105%) (Table 3.7). An important upward trend in the rainfall for August is revealed by the monthly trend study (Table 3.7). The study showed that despite large variations in rainfall amounts over the study period (Table 3.6), there has been no significant change in the percentage share of rainfall on the monthly scale (Table 3.7).

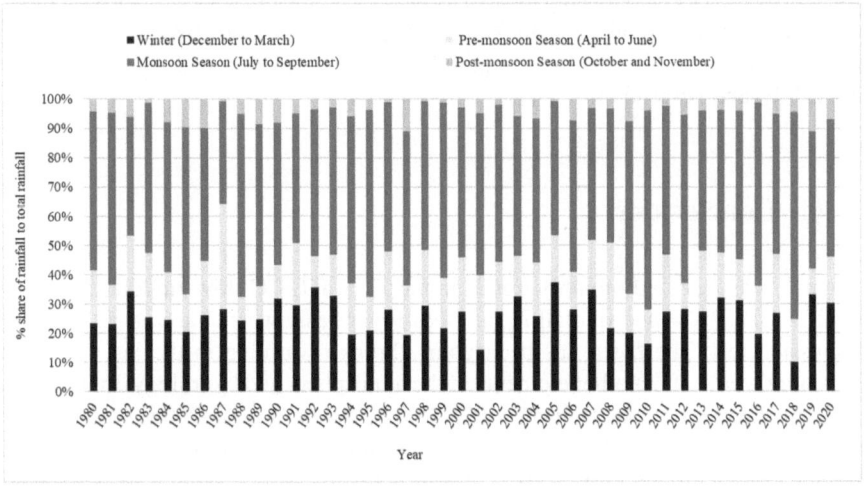

Fig. 3.22 Time series of % share of total rainfall of the Beas basin

Table 3.7 Descriptive statistics and trend of % share of total rainfall of the Beas basin

Time scale	Mean (%)	Std. deviation (%)	Coefficient of variation (%)	Trend (%/decade)
Jan	6.27	3.53	56.35	0.58
Feb	9.10	4.29	47.10	0.39
Mar	7.93	3.80	47.97	−0.77
Apr	5.19	2.76	53.29	−0.50
May	3.90	2.75	70.50	−0.45
Jun	7.78	4.10	52.72	0.47
Jul	24.16	7.01	28.99	−1.46
Aug	20.75	5.19	24.98	1.27*
Sep	8.12	4.38	53.88	0.86
Oct	1.86	2.00	107.90	−0.15
Nov	1.80	1.80	100.27	0.17
Dec	3.14	2.26	71.92	−0.35
Winter	26.34	6.18	23.47	−0.12
Pre-monsoon	16.83	5.10	30.30	−0.40
Monsoon	53.02	7.51	14.16	0.54
Post-monsoon	4.93	3.00	60.96	−0.16

Sig at 0.10*, 0.05**, and 0.01***

3.9 Relationship of Air Temperature with Other Climate Variables

Correlation coefficients (r) result shows the level of association of T_{mean} with other variables in the study area (Table 3.8). ET has a significant positive relation with T_{mean} in February and July, while it shares negative relationship with T_{mean} in April and May. Low cloud cover shows a negative relationship with T_{mean} in February–May, July, and December, while it has a negative relationship with T_{mean} in September. CBH has negative relation with T_{mean} in September and positive relationship in February–April and December. Rainfall has negative relation with T_{mean} in March–May, July, and November.

3.10 Conclusions

Annual, seasonal, and monthly trends in the T_{mean}, ET, CC (low, middle, high, and total), CBH, PWV, and rainfall in the Beas River basin were analyzed by the OLS regression test. The basin has been going through warming in the last 40 years. Warming would affect other elements of the atmosphere and hydrology. In the upper parts of the Beas basin, Bahang (1974–2005), Solang (1982–2005), and Dhundi (1989–2005) all showed a substantial downward trend in snowfall, according to research by Bhutiyani et al. (2010). Additionally, the study found a negative relationship between winter T_{mean} and total snowfall accumulations at the sites between

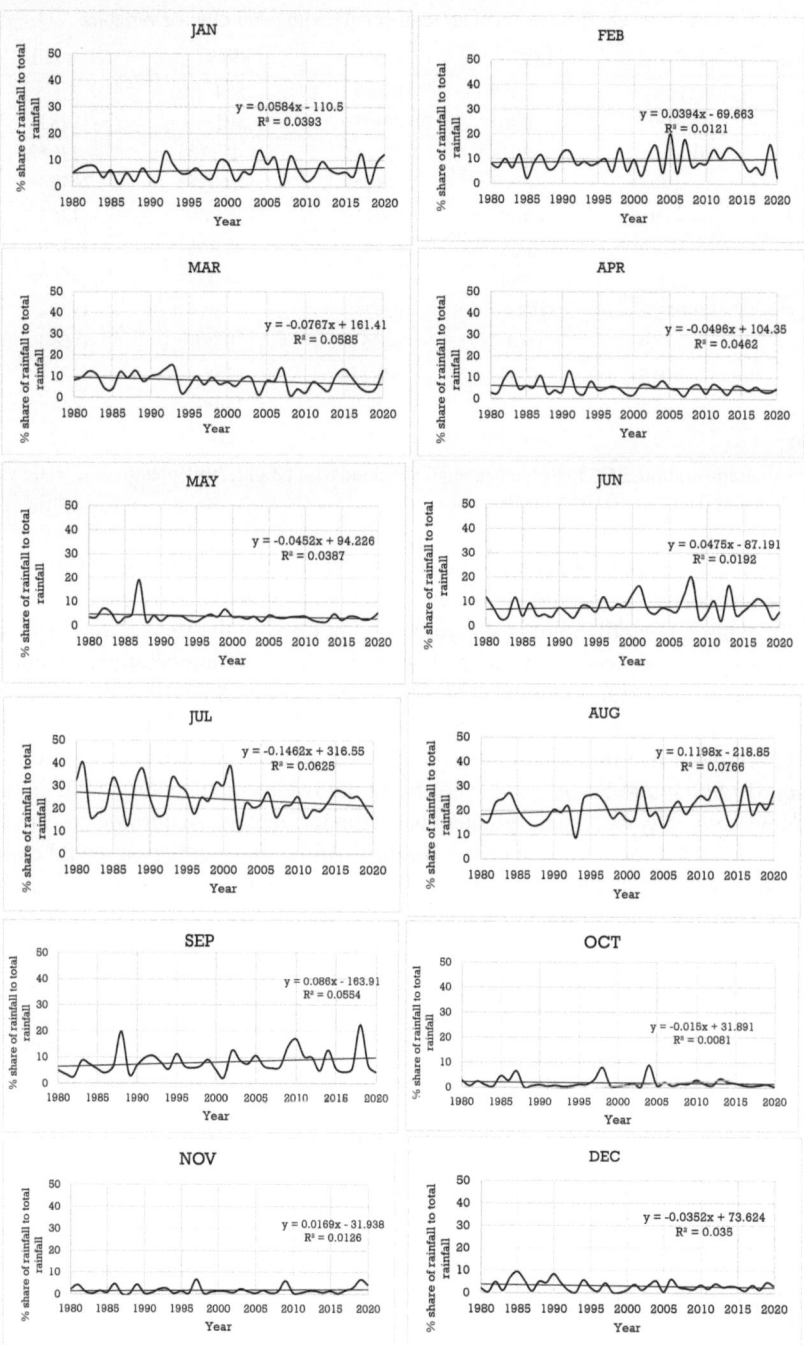

Fig. 3.23 Time series of % share of total rainfall of the Beas basin

Table 3.8 Correlation coefficients (r) of air temperature with other climate variables

Time	ET	LCC	CBH	PWV	Rainfall
Jan	0.165	−0.058	0.305	.554**	0
Feb	.537**	−.486**	.790**	.501**	−0.281
Mar	0.11	−.802**	.854**	−0.118	−.720**
Apr	−.497**	−.696**	.755**	−.495**	−.657**
May	−.419**	−.405**	0.087	−0.189	−.735**
Jun	−0.162	−0.21	0.23	−0.305	−0.149
Jul	.320*	−.384*	0.291	−0.073	−.477**
Aug	0.023	0.011	0.02	.378*	−0.094
Sep	−0.169	.319*	−.451**	.585**	−0.063
Oct	−0.018	−0.05	−0.259	.318*	−0.224
Nov	0.034	−0.245	0.241	.354*	−.414**
Dec	0.178	−.429**	.418**	0.26	−0.22

ET evapotranspiration, *LCC* low cloud cover, *CBH* cloud base height, *PWV* precipitable water vapor
Significant at 0.05* and 0.01** levels

1991 and 2005. In the upper Beas basin between 2000 and 2010, Rani (2014) discovered a negative association between the snow cover area and mean T_{min} throughout the winter. Overall, there is a slight warming trend in the Beas River basin, and changes in T_{mean} are consistent with changes in global air temperature, though the magnitude is not great. Forsythe et al. (2014) predict that the UIB's mean temperature will increase year-round (annual mean +4.8 °C) (2071–2100). Wijngaard et al. (2017) found a rise in the amplitude of climatic means and extremes in the Indo-Ganga-Brahmaputra basin toward the end of the twenty-first century, with climatic extremes typically becoming greater than climatic means. According to Shrestha et al. (2015), the rough terrain of the Hindu Kush Himalayan region would experience an average temperature increase of 1–2 °C (and in certain places as high as 4–5 °C) by 2050. According to Baig et al. 2021, every season in the UIB will experience temperature increases in the future (2075–1999), with September and October experiencing the largest increases at roughly 8 °C. However, slight changes in air temperature in a region like the Himalayas could have a long-term impact on the availability of water resources because rising air temperatures increase evaporation rates and alter the type of precipitation, which may negatively impact streamflow. ET in the basin showed both increasing and decreasing trends during the study period. A similar trend is also observed in cloud cover conditions of the area. CBH is rising in the area particularly in April during the study period. PWV is rising in the area during the pre-monsoon months. However, the amount of rainfall is declining in post-monsoon months. Future precipitation in the UIB is anticipated to increase year-round (highest seasonal mean change: +27%, annual mean change: +18%) with higher intensity in the wettest months (February–April during 2071–2100) (Forsythe et al. 2014). According to Baig et al. (2021), future precipitation (2075–2099) will increase every month across the UIB, with the increase in the period between July and August being particularly notable. By the end of the century, Shirsat et al. (2021) predicted that the upper Beas basin's mean annual temperature will rise by 3 °C (RCP4.5) and 5.6 °C (RCP8.5). By the end of the century, precipitation may rise between 9.9% (RCP4.5) and 19.5% (RCP8.5). The percentage of snowfall in total precipitation will steadily decrease, going from 59.6% in the baseline to 48.3% (RCP4.5) and 45% (RCP8.5) by 2050; it will then decrease even further, going from 41.6% (RCP4.5) to 33.7% (RCP8.5) by 2090. The expected rise in actual ET in the basin ranges from 23.3% (2050 RCP4.5) to 52.7% (2090 RCP8.5). Rainfall that varies over time may have an impact on hydropower production and other operations that depend directly or indirectly on water availability. In the period between 1962 and 1995, Bhutiyani et al. (2008) revealed a considerable decrease in the Beas River discharge near Thalout throughout the spring and monsoon. As a result, more extensive climate monitoring is required throughout the Beas River watershed for effective planning and management.

References

Ali SH, Shafqat MN, Eqani SA, Shah ST (2019) Trends of climate change in the upper Indus basin region, Pakistan: implications for cryosphere. Environ Monit Assess 191(2):1–2

Babar SF, Ramesh H (2013) Analysis of southwest monsoon rainfall trend using statistical techniques over Nethravathi basin. Int J Civ Eng Technol 2(3):130–136

Baig S, Sayama T, Yamada M (2021) Impacts of climate change on river flows in the upper Indus basin and its subbasins. https://doi.org/10.21203/rs.3.rs-404691/v1

Bengtsson L, Hagemann S, Hodges KI (2004) Can climate trends be calculated from reanalysis data? J Geophys Res Atmos 109(D11)

Bhutiyani MR, Kale VS, Pawar NJ (2008) Changing streamflow patterns in the rivers of northwestern Himalaya: implications of global warming in the 20th century. Curr Sci 10:618–626

Bhutiyani M, Kale V, Pawar N (2010) Climate change and the precipitation variations in the northwestern Himalaya: 1866–2006. Int J Climatol 30(4):535–548

Census of India (2011) Provisional population totals Paper 1 of 2011: Himachal Pradesh. Available via http://www.censusindia.gov.in/2011-prov-results/prov_data_products_himachal.html. Accessed 20 May 2021

Dimri AP, Kumar D, Choudhary A, Maharana P (2018) Future changes over the Himalayas: maximum and minimum temperature. Glob Planet Chang 162:212–234

Forsythe N, Fowler HJ, Blenkinsop S, Burton A, Kilsby CG, Archer DR, Harpham C, Hashmi MZ (2014) Application of a stochastic weather generator to assess climate change impacts in a semi-arid climate: the Upper Indus Basin. J Hydrol 517:1019–1034

Forsythe N, Fowler HJ, Li X-F, Blenkinsop S, Pritchard D (2017) Karakoram temperature and glacial melt driven by regional atmospheric circulation variability. Nat Clim Chang 7(9). https://doi.org/10.1038/nclimate3361

Gupta A, Dimri AP, Thayyen R, Jain S, Jain S (2020) Meteorological trends over Satluj River Basin in Indian Himalaya under climate change scenarios. J Earth Syst Sci 129:161. https://doi.org/10.1007/s12040-020-01424-x

Hassan M, Du P, Mahmood R, Jia S, Iqbal W (2019) Streamflow response to projected climate changes in the Northwestern Upper Indus Basin based on regional climate model (RegCM4. 3) simulation. J Hydro-Environ 27:32–49

Held IM, Soden BJ (2000) Water vapor feedback and global warming. Annu Rev Environ Resour 25(1):441–475

Huss M, Hock R (2018) Global-scale hydrological response to future glacier mass loss. Nat Clim Chang 8(2):135–140

Hussain D, Kao HC, Khan AA, Lan WH, Imani M, Lee CM, Kuo CY (2020) Spatial and temporal variations of terrestrial water storage in upper Indus basin using GRACE and altimetry data. IEEE Access 8:65327–65339

Hussain A, Cao J, Hussain I, Begum S, Akhtar M, Wu X, Guan Y, Zhou J (2021) Observed trends and variability of temperature and precipitation and their global teleconnections in the upper Indus Basin, Hindukush-Karakoram-Himalaya. Atmosphere 12(8):973

India Meteorological Department (IMD) (2019) Annual climate summary. Available via https:// imdpune.gov.in/Links/annual_summary_2019.pdf. Accessed 1 Feb 2021

IPCC (2021) Climate change widespread, rapid, and intensifying – IPCC. IPCC press release AR6, pp 1–6

Jaswal AK, Kore PA, Singh V (2017) Variability and trends in low cloud cover over India during 1961–2010. Mausam 68(2):235–252

Krakauer NY, Lakhankar T, Dars GH (2019) Precipitation trends over the Indus basin. Climate 7(10):116

Krishnan R, Shrestha AB, Ren G, Rajbhandari R, Saeed S, Sanjay J, Syed M, Vellore R, Xu Y, You Q, Ren Y (2019) Unravelling climate change in the Hindu Kush Himalaya: rapid warming in the mountains and increasing extremes. In: The Hindu Kush Himalaya assessment. Springer, Cham, pp 57–97

Kumar V, Jain SK, Singh Y (2010) Analysis of long-term rainfall trends in India. Hydrol Sci J 55(4):484–496

Kumar N, Tischbein B, Beg MK (2018) Multiple trend analysis of rainfall and temperature for a monsoon-dominated catchment in India. Meteorol Atmos Phys 131(4):1019–1033

Liaqat MU, Grossi G, Ranzi R (2022) Characterization of interannual and seasonal variability of hydro-climatic trends in the Upper Indus Basin. Theor Appl Climatol 147(3):1163–1184

Liu X, Cheng Z, Yan L, Yin ZY (2009) Elevation dependency of recent and future minimum surface air temperature trends in the Tibetan Plateau and its surroundings. Glob Planet Chang 68:164–174

Mayewski PA, Perry LB, Matthews T, Birkel SD (2020) Climate change in the Hindu Kush Himalayas: basis and gaps. One Earth 3(5):551–555. https://doi.org/10.1016/j. oneear.2020.10.007

Nie Y, Liu Q, Wang J, Zhang Y, Sheng Y, Liu S (2018) An inventory of historical glacial lake outburst floods in the Himalayas based on remote sensing observations and geomorphological analysis. Geomorphology 308:91–106

Nie Y, Pritchard HD, Liu Q, Hennig T, Wang W, Wang X, Liu S, Nepal S, Samyn D, Hewitt K, Chen X (2021) Glacial change and hydrological implications in the Himalaya and Karakoram. Nat Rev Earth Environ 2(2):91–106

Nüsser M, Dame J, Parveen S, Kraus B, Baghel R, Schmidt S (2019) Cryosphere-fed irrigation networks in the northwestern Himalaya: precarious livelihoods and adaptation strategies under the impact of climate change. Mt Res Dev 39(2):R1–R1

Rani S (2014) Assessment of the influence of climate variability on the snow cover area of the upper Beas River basin. Unpublished M. Phil. Dissertation, Centre for the Study of Regional Development, Jawaharlal Nehru University, New Delhi

Rani S, Sreekesh S (2018) Variability of temperature and rainfall in the upper Beas basin, Western Himalayas. In: Mal S, Singh RB, Huggel C (eds) Climate change, extreme events and disaster risk reduction, Sustainable development goals series. Springer, Cham, pp 101–120

Rani S, Kumar R, Maharana P (2022) Climate change, its impacts, and sustainability issues in the Indian Himalaya: an introduction. In: Climate change. Springer, Cham, pp 1–27

Ray LK, Goel NK, Arora M (2019) Trend analysis and change point detection of temperature over parts of India. Theor Appl Climatol 138(1–2):153–167

Ren YY, Ren GY, Sun XB, Shrestha AB, You QL, Zhan YJ, Rajbhandari R, Zhang PF, Wen KM (2017) Observed changes in surface air temperature and precipitation in the Hindu

Kush Himalayan region over the last 100-plus years. Adv Clim Chang Res 8(3). https://doi. org/10.1016/j.accre.2017.08.001

Saxena R, Mathur P (2019) Recent trends in rainfall and temperature over North West India during 1871–2016. Theor Appl Climatol 135(3–4):1323–1338

Schickhoff U, Singh RB, Mal S (2016) Climate change and dynamics of glaciers and vegetation in the Himalaya: an overview. In: Singh R, Schickhoff U, Mal S (eds) Climate change, glacier response, and vegetation dynamics in the Himalaya. Springer, Cham. https://doi. org/10.1007/978-3-319-28977-9_1

Schmidt S, Nüsser M (2012) Changes of high altitude glaciers from 1969 to 2010 in the Trans-Himalayan Kang Yatze Massif, Ladakh, northwest India. AAAR 44(1):107–121

Schmidt S, Nüsser M (2017) Changes of high altitude glaciers in the Trans-Himalaya of Ladakh over the past five decades (1969–2016). Geosciences 7(2):27

Schmidt S, Nüsser M, Baghel R, Dame J (2020) Cryosphere hazards in Ladakh: the 2014 Gya glacial lake outburst flood and its implications for risk assessment. Nat Hazards 104(3):2071–2095

Shirsat TS, Kulkarni AV, Momblanch A, Randhawa SS, Holman IP (2021) Towards climate-adaptive development of small hydropower projects in Himalaya: A multi-model assessment in upper Beas basin. J Hydrol Reg 34:100797

Shrestha AB, Agrawal NK, Alfthan B, Bajracharya SR, Maréchal J, van Oort B (eds) (2015) The Himalayan Climate and Water Atlas: impact of climate change on water resources in five of Asia's major river basins. ICIMOD, GRID-Arendal and CICERO

Singh RB, Mal S (2014) Trends and variability of monsoon and other rainfall seasons in Western Himalaya, India. Sci Lett 15(3):218–226. https://doi.org/10.1002/asl2.494

Wagner T, Beirle S, Grzegorski M, Platt U (2006) Global trends (1996–2003) of total column precipitable water observed by Global Ozone Monitoring Experiment (GOME) on ERS-2 and their relation to near-surface temperature. J Geophys Res Atmos 111(D12)

Wang P, Yamanaka T, Qiu GY (2012) Causes of decreased reference evapotranspiration and pan evaporation in the Jinghe River catchment, northern China. Environmentalist 32(1):1

Wijngaard RR, Lutz AF, Nepal S, Khanal S, Pradhananga S, Shrestha AB, Immerzeel WW (2017) Future changes in hydro-climatic extremes in the Upper Indus, Ganges, and Brahmaputra River basins. PLoS One 12(12):e0190224

Xenarios S, Gafurov A, Schmidt-Vogt D, Sehring J, Manandhar S, Hergarten C, Shigaeva J, Foggin M (2019) Climate change and adaptation of mountain societies in Central Asia: un-certainties, knowledge gaps, and data constraints. Reg Environ Chang 19(5):1339–1352. https://doi. org/10.1007/s10113-018-1384-9

You QL, Ren GY, Zhang YQ, Ren YY, Sun XB, Zhan YJ, Shrestha AB, Krishnan R (2017) An overview of studies of observed climate change in the Hindu Kush Himalayan (HKH) region. Adv Clim Chang Res 8(3):141–147. https://doi.org/10.1016/j.accre.2017.04.001

Zhang L, Wu L, Gan B (2013) Modes and mechanisms of global water vapor variability over the twentieth century. J Clim 26(15):5578–5593

Chapter 4
Land Use/Land Cover: Status and Changes

Abstract Streamflow of any area is influenced by land use/land cover (LULC) distribution. If LULC changes are not taking place in accordance with the environmental conditions, they present significant issues and barriers to the sustainable development of any location. With an increasing population comes a greater need for food and energy. As a result, human activities such as farming, hydropower project construction, and forest clearance become more intense with time, resulting in LULC changes. The results demonstrate that, except for cultivated land and barren land, all LULC classes are expanding. Forest regions and water bodies have absorbed most of the surge. Moreover, a modest rise was noted in the study region's built-up area. As compared to very low-density rural areas, it demonstrates that dense urban areas are growing more rapidly than other classes, pointing to a rise in urbanization. Snow cover area (SCA) is increasing in all months except January, April, September, and October from 2001 to 2020. While the trend is minor, a decline in SCA during the pre-monsoon season might increase discharge, but over time, it may cause a decline in streamflow. Hence, it requires further analysis using fine-resolution data.

Keywords Land use/land cover · Urbanization · Settlement density · Snow cover area

4.1 Introduction

Land use/land cover (LULC) is another causative factor that has an impact on streamflow. If LULC changes are not occurring in line with the environmental circumstances, they pose serious problems and obstacles to the sustainable development of any place. Demand for food and energy rises as the population grows. As a result, human activities like farming, building hydroelectric projects, clearing forests, etc. intensify over time and result in LULC changes. Changes in LULC affect how various ecosystem components interact with one another and can occasionally have negative consequences including landslides, floods, and soil erosion. Not all

S. Rani, *Climate, Land-Use Change and Hydrology of the Beas River Basin,
Western Himalayas*, Advances in Asian Human-Environmental Research,
https://doi.org/10.1007/978-3-031-29525-6_4

changes in the LULC are the result of human activity; occasionally, natural events like forest fires, landslides, and flash floods also affect the land cover. Hence, monitoring and controlling environmental changes and mitigating climate change now heavily rely on spatiotemporal analysis of LULC. Studies from throughout the world evaluated how LULC has changed for managing water resources (Prasena and Shrestha 2013; Birhanu et al. 2019; Awotwi et al. 2019; Woyessa and Welderufael 2021). Additionally, research is conducted on Indian river basins to comprehend how variations in LULC affect the watershed hydrology (Anand et al. 2021; Gaur et al. 2021; Naha et al. 2021). It is crucial to understand LULC to calculate the impact of these evolving patterns on streamflow in any region. The Beas basin's LULC trend from 1975 to 2020 is depicted in this chapter both temporally and spatially. Various LULC maps of the Beas basin are shown in the first part, along with an indication of their degree of accuracy. The LULC changes in the study area over the period of analysis are discussed in the second part. The third part examines the snow cover area (SCA) of the basin between 2001 and 2020. The chapter is summarized in the last section.

4.2   Accuracy of LULC

The findings of the accuracy evaluation of the classified images of the upper Beas basin (up to Pandoh Dam) indicate that the overall accuracy of the classified images ranged between 83% (1980) and 85% (2000), while the Kappa coefficients varied from 0.78 to 0.86. (Table 4.1). Across all classes in 1980, producers' built-up area accuracy is quite poor. It was taken from toposheets because there weren't many major built-up places in 1980. But as time went on, the built-up area grew and covered more ground. In images, these areas are modest and interspersed with cultivated and BUW areas. Additionally, there are fewer ground truth points accessible for the class, which leads to a lower producer's accuracy and a greater user's accuracy.

Table 4.1 Accuracy assessment (%) of LULC maps of the upper Beas basin

LULC classes	UA 1980	PA 1980	UA 2000	PA 2000	UA 2020	PA 2020
Built-up area	57	40	95	72	96	71
Agriculture	76	76	82	94	63	71
Forest	76	94	81	100	82	78
Grassland	86	75	93	93	80	98
BUW barren/unculturable/wasteland	82	94	78	88	78	87
Water bodies	100	71	100	89	100	91
Snow	95	81	96	84	96	93
Overall accuracy		83		88		85
Kappa coefficient		0.78		0.86		0.82

Note: *UA* user's accuracy, *PA* producer's accuracy

4.3 Land Use/Land Cover Status and Changes

The spatial distribution of the entire Beas basin has been computed from MODIS MOD12Q1 data of 2001–2020 given in Chap. 1. The results indicate that open forest and cropland covered a majority of the study area followed by dense forest in both years (Table 4.2 and Fig. 4.1). Glacier/snow and water bodies together covered less than 2% and 3% of the total area of the basin in 2001 and 2020, respectively. Dense forests, barren land, and natural herbaceous are dominant in the upper Beas basin up to Pandoh Dam (Fig. 4.1). Central part of the Beas basin is dominated by open forest, while the lower areas are mostly covered by cropland. Cropland is spreading faster in the lower part of the basin including the Punjab plains which is

Table 4.2 Status and changes in LULC of the study area

LULC classes	2001 Area (km²)	2020 Area (km²)	Change (2001–2020) Area (km²)	Change (2001–2020) Area (%)
Barren	1362	1210	−152	−11
Permanent snow and ice	140	275	135	96
Water bodies	122	149	27	22
Urban and built-up lands	74	78	4	6
Dense forests	1624	2190	566	35
Open forests	7124	7798	674	9
Forest/cropland mosaics	500	647	147	29
Natural herbaceous	2229	2054	−176	−8
Croplands	6367	5136	−1231	−19
Shrublands	0.43	5	4.9	1150

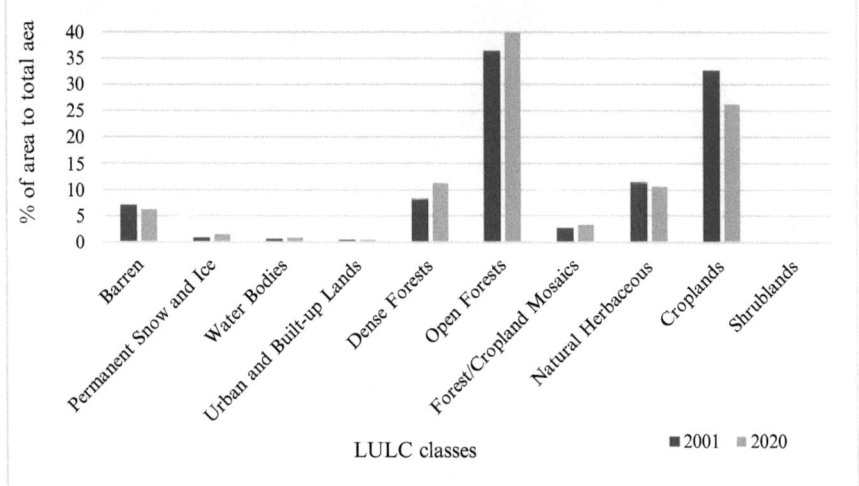

Fig. 4.1 Status of LULC of the study area

also the state of the green revolution. It indicates that among surface features, forest, barren land, and cropland are the most influencing factors of surface flow (Fig. 4.2).

As the built-up area covered less than 3% part of the Beas basin, MODIS MOD12Q1 data doesn't capture it correctly. Its spatial and temporal pattern has been shown by high-resolution product of the Global Human Settlement Layer Built-up area grid (GHS-BUILT) derived from Landsat (Chapter 1 material and method section) during 1975–2014 (Fig. 4.3a). Built-up area was about 0.38% of

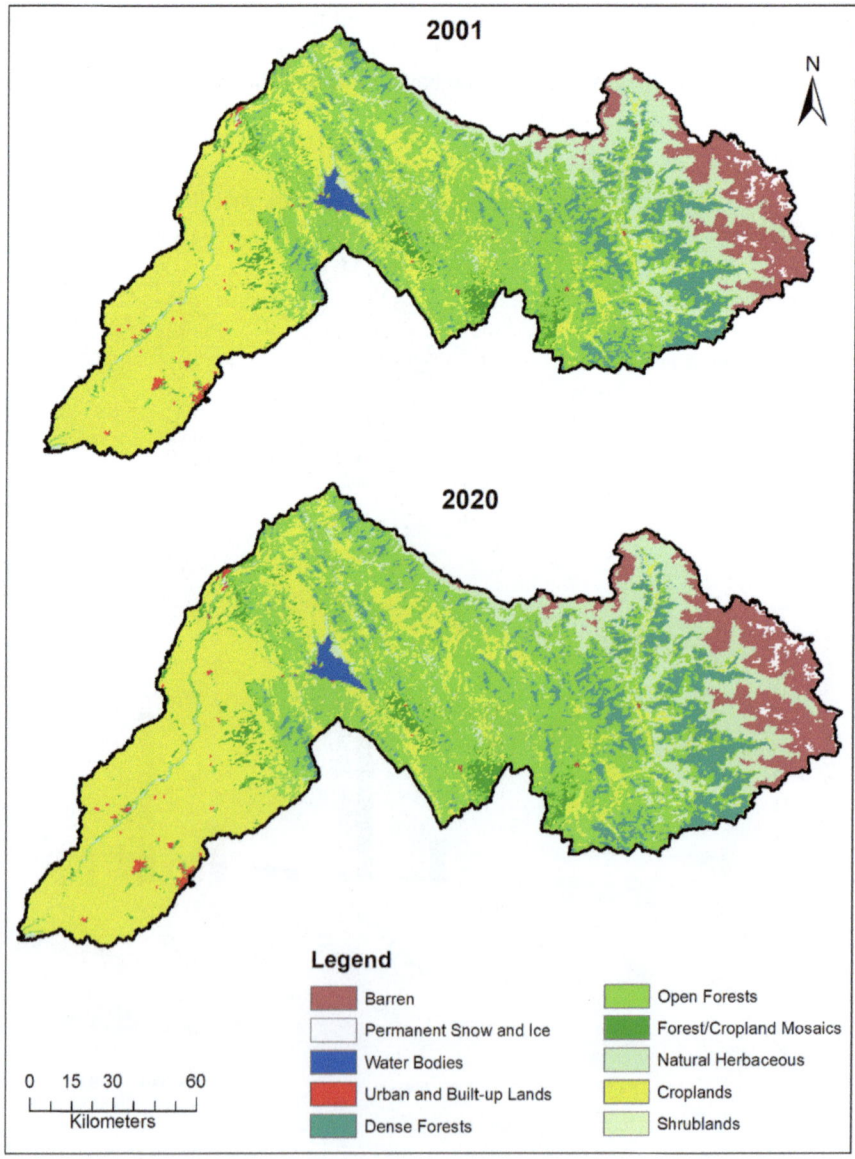

Fig. 4.2 Spatial distribution of LULC classes of the Beas River basin

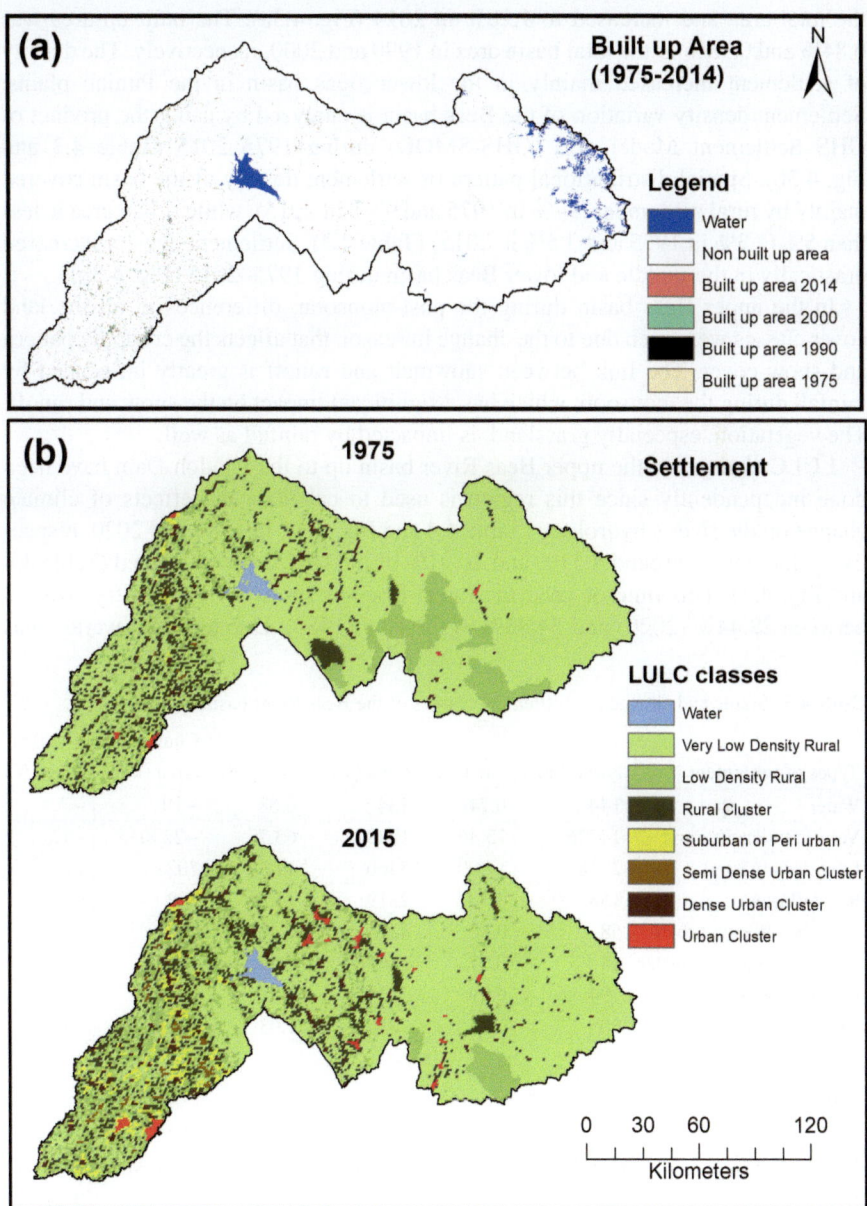

Fig. 4.3 (**a**) Temporal distribution of built-up area of the Beas basin; (**b**) spatial distribution of settlement classes of the Beas River basin

the total area and increased to 1.28% in 2014 (Fig. 4.3a). The built-up area was 0.84% and 0.93% of the total basin area in 1990 and 2000, respectively. The density of settlement increased mainly in the lower Beas basin in the Punjab plains. Settlement density variation of the Beas basin is analyzed by using the product of GHS Settlement Model grid (GHS-SMOD) during 1975–2015 (Table 4.3 and Fig. 4.3b). Spatial distributional pattern of settlement density of the basin covered mainly by rural settlement (97% in 1975 and 93% in 2015), while urban area is less than 5% (1.3% in 1975 and 3.5% in 2015) (Table 4.3). Settlement density increased drastically in the middle and lower Beas basin during 1975–2015 (Fig. 4.3b).

In the upper Beas basin during the post-monsoon, differences in all the land cover classes were seen due to the change in season that affects the cropping pattern and snow cover. The link between snowmelt and runoff is greatly influenced by rainfall during the monsoon, which has a significant impact on the snow and runoff. The vegetation, especially grassland, is impacted by rainfall as well.

LULC changes in the upper Beas River basin up to the Pandoh Dam have been done independently since this region is used to calculate the effects of climate change on the river's hydrology (Table 4.4 and Fig. 4.4). In 1980 and 2020, respectively, there were around 4.51% and 10.91% of the land that was farmed (Table 4.4 and Fig. 4.4). This time of year, the forest loses its snow cover. Forestry covered between 29.44% (2020) and 34.43% of the land (1980). Due to rain, several areas

Table 4.3 Status and changes in settlement classes of the Beas River basin

Types of settlement	1975		2015		Change (1975–2015)	
	Area (km²)	Area (%)	Area (km²)	Area (%)	Area (km²)	Area (%)
Water	144	0.74	134	0.68	−10	−7
Very-low-density rural	14776	75.46	12486	63.77	−2290	−15
Low-density rural	2758	14.09	3460	17.67	702	25
Rural cluster	1587	8.10	2319	11.84	732	46
Suburban or peri-urban	58	0.30	482	2.46	424	731
Semi-dense urban cluster	14	0.07	116	0.59	102	729
Dense urban cluster	194	0.99	382	1.95	188	97
Urban cluster	50	0.26	202	1.03	152	304

Table 4.4 Status of LULC of the study area

LULC classes	1980		2000		2020	
	Area (km²)	Area (%)	Area (km²)	Area (%)	Area (km²)	Area (%)
Built-up area	7	0.13	19	0.34	32	0.60
Cultivated land	243	4.51	305	5.67	587	10.91
Forest	1853	34.43	1824	33.89	1585	29.44
Grassland	114	2.11	594	11.03	704	13.08
BUW: barren/ unculturable/wasteland	1975	36.69	1976	36.71	1789	33.23
Water bodies	12	0.22	5	0.10	9	0.16
Snow	1179	21.91	659	12.25	676	12.57

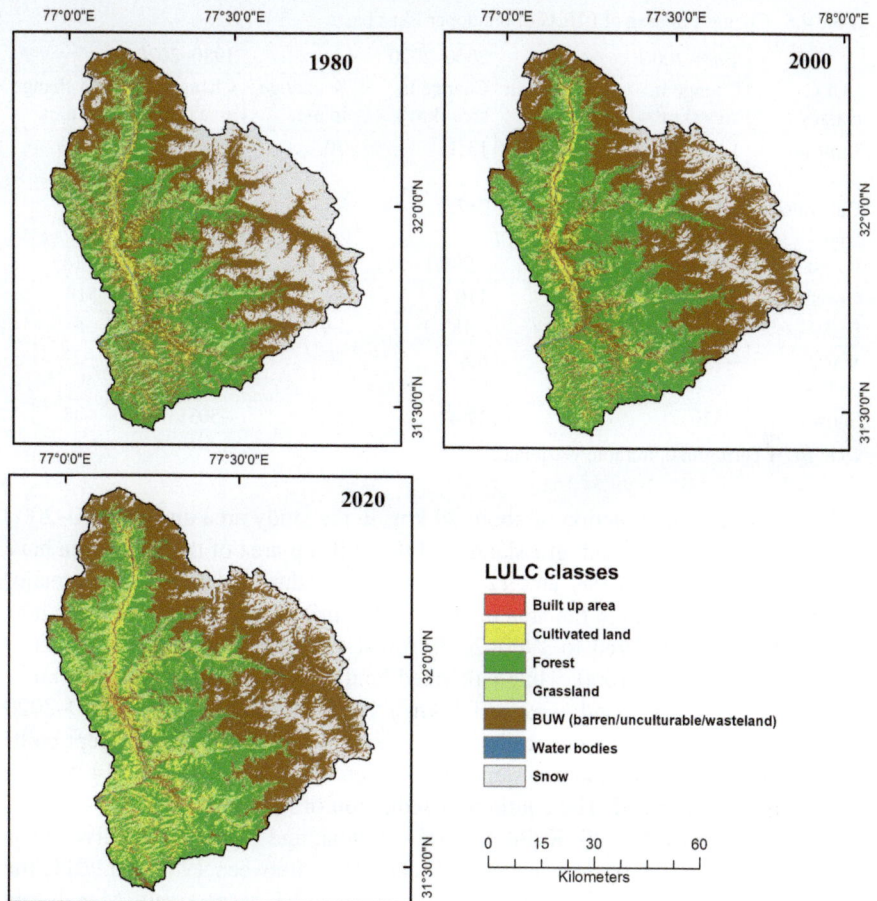

Fig. 4.4 Spatial distribution of LULC classes of the upper Beas River basin

of BUW during this season become grassland. In 1980 and 2020, the proportion of grassland ranged from 2.11% to 13%. Due to the little snow cover this season, BUW's extent has grown. Throughout the time of the study, its range stayed at more than 33% in the basin. Because this season saw the basin's lowest rainfall (29–35 mm), water bodies are represented to have an area below 0.22%. During this season, snow accumulation also begins in late October and reaches a peak of around 19% (Rani 2014).

Table 4.5 shows the variations in LULC classes that took place in the upper Beas basin during the duration of the study. Over the period 1980–2020, there were two major trends in LULC: growth in a built-up area (70 km^2) and cultivated land (344 km^2) and a decline in the forest (268 km^2) (Table 4.5). The water body under observation had a little alteration of roughly 3 km^2. Snowfall decreased over the period as well, although it is hard to draw any conclusions about a pattern based on data from only two time periods.

Table 4.5 Changes in area of LULC of the upper Beas basin

LULC classes	1980–2000		2000–2020		1980–2020	
	Change in area (km²)	% change in area	Change in area (km²)	% change in area	Change in area (km²)	% change in area
Built-up area	12	171.4	13.3	70	25	362
Cultivated land	62	25.5	282.2	93	344	142
Forest	−29	−1.6	−239.1	−13	−268	−14
Grassland	480	421.1	110.3	19	590	518
BUW	1	0.1	−187.3	−9	−186	−9
Water bodies	−7	−58.3	3.8	76	−3	−27
Snow	−520	−44.1	17.4	3	−503	−43

Note: *BUW* barren/unculturable/wasteland

Forest showed a reduction of about 29 km² in the study area during 1980–2000 (Table 4.5). Cultivated land, grassland, and the built-up area of the study area have increased by 25.5%, 421%, and 12%, respectively, during 1980–2000. A major reduction was observed in the area of snow (520 km²) in the upper Beas basin during the period. Compared to 1980–2000, forest (239 km²) and BUW (187 km²) showed a reduction in 2000, while cultivated land (289 km²), grassland (110 km²), and snow (17.4 km²) and water (3.8 km²) have increased during 2000–2020 (Table 4.5). All land cover classes have decreased during 2000–2020, except built-up area, cultivated land, and grassland. These were increased by 362%, 142%, and 518% during this period. The continuous reduction of forests during 1980–2020 is a matter of concern. Manali, Kullu, Samshi, Bhuntar, and Banjar are the five towns that are located in the basin (Census of India 2011). Between 1981 and 2011, the growth rates of Manali and Bhuntar were 252% and 62%, respectively. Manali saw an exceptional growth rate of 157.50% from 1991 to 2001, the highest among Himachal Pradesh's metropolitan centers. This indicates that Manali attracted a sizable population during this time. In the same decade, Bhuntar's population increased significantly by 43.3%. As a result, throughout time, the built-up area has risen at both places due to growth in the residential area, commercial spaces, transportation network, etc. The whole Kullu district is covered by the upper Beas basin (excluding Ani and Herman block), which ranks fourth in terms of decadal population increase (2001–2011) with 14.8% more people than the state average of 12.9%. The Kullu district's economy is mostly dependent on agriculture. It ranks fourth among the districts' working population in the state with 197,141 cultivators. As a result, cultivated land outpaces other types of land cover.

To comprehend how LULC converts in various land classifications, a transition matrix was created (Table 4.6 and Fig. 4.5). It would provide a better estimate for the conversion of LULC. During 1980–2020 (Table 4.6), 36.3%, 17.2%, 11.3%, 27.8%, and 5.3% of the built-up area are converted to cultivated land, forest, grassland, BUW, and water bodies, respectively. A total of 35.6% of cultivated land has been turned into grassland, 22.5% into desert, and 0.3% into water bodies. A total

Table 4.6 Transition matrix (% of area) of LULC of the upper Beas basin

	1980						
2020 LULC classes	Built-up area	Cultivated land	Forest	Grassland	BUW	Water bodies	Snow
Built-up area	2.1	0.2	0.1	0.1	0.0	3.0	0.0
Cultivated land	36.3	24.1	1.2	4.4	1.5	13.7	0.0
Forest	17.2	17.3	78.1	13.4	1.3	26.3	0.1
Grassland	11.3	35.6	9.3	61.5	4.0	1.7	0.0
BUW	27.8	22.5	11.0	20.6	59.7	16.7	13.6
Water bodies	5.3	0.3	0.1	0.0	0.1	38.5	0.0
Snow	0.0	0.0	0.0	0.0	33.3	0.1	86.3
Class total	100	100	100	100	100	100	100

of 1.2% of forested land has been changed to agriculture, 9.3% to grassland, 11% to BUW, and 0.8% to water bodies. About 4.4% of the area of grassland has been changed into arable land, 13.3% into woodland, and 20.6% into cultivated land. A little over 3% of water bodies have been turned into built-up areas, 13.3% into cultivated land, and 16.7% into desert. And 13.6% of the snow has turned into bare soil.

4.4   Status of Snow Cover Area

SCA is a significant component that affects streamflow among all of the basin's land cover types. Therefore, it is crucial to evaluate the seasonal variation in SCA in order to comprehend its spatial pattern in the basin. Snow precipitation is limited to higher altitudes of the Beas River. Hence, SCA analysis has been done in the upper Beas basin using MOD10A2 datasets. Based on SCA evaluation, the mean monthly SCA for the period 2001–2020 indicates that snow accumulates from late October to mid-March and then decreases from April to the end of September (Figs. 4.6 and 4.7). The highest and lowest SCA were noted in February and September months of the study period.

SCA is a more dynamic land cover compared to other land cover categories. Mean monthly variation in SCA indicates the accumulation and ablation of snow under different elevation zones in the upper Beas basin during 2001–2020. SCA covers the majority of the area >3000-m above the mean sea level (AMSL). For all the months in the study area, the distribution of SCA was evaluated based on elevation. Snow coverage reaches a height of 2000 meters in January and stays at or above 4000 meters in August (Fig. 4.8).

The trend in the SCA of the study area doesn't indicate a significant trend in the study area. However, it indicates a declining trend in pre-monsoon during the study period (Table 4.7). Rani (2014) investigated the seasonal and inter-annual fluctuation in SCA in the upper Beas basin between 2000 and 2010. During the winter and pre-monsoon season, the study observed a decreasing pattern in SCA. Additionally,

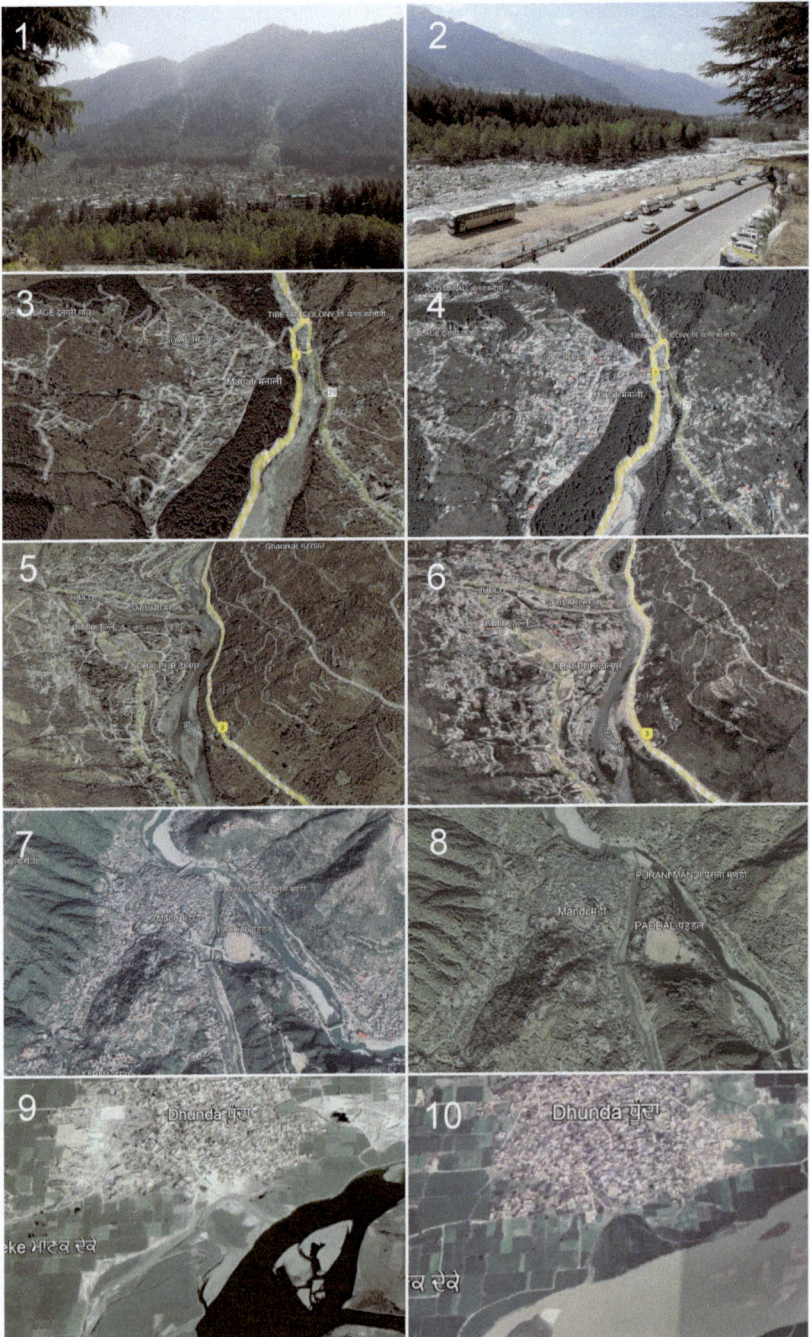

Fig. 4.5 Spatial distribution of urban growth over other LULC classes of the Beas River basin. (1, 2) Manali town expansion. Remaining images show the urban growth of selected towns between 2004 (left panel) and 2021 (right panel) based on Google Earth images: (3, 4) Manali; (5, 6) Kullu; (7, 8) Mandi; (9, 10) Dhunda

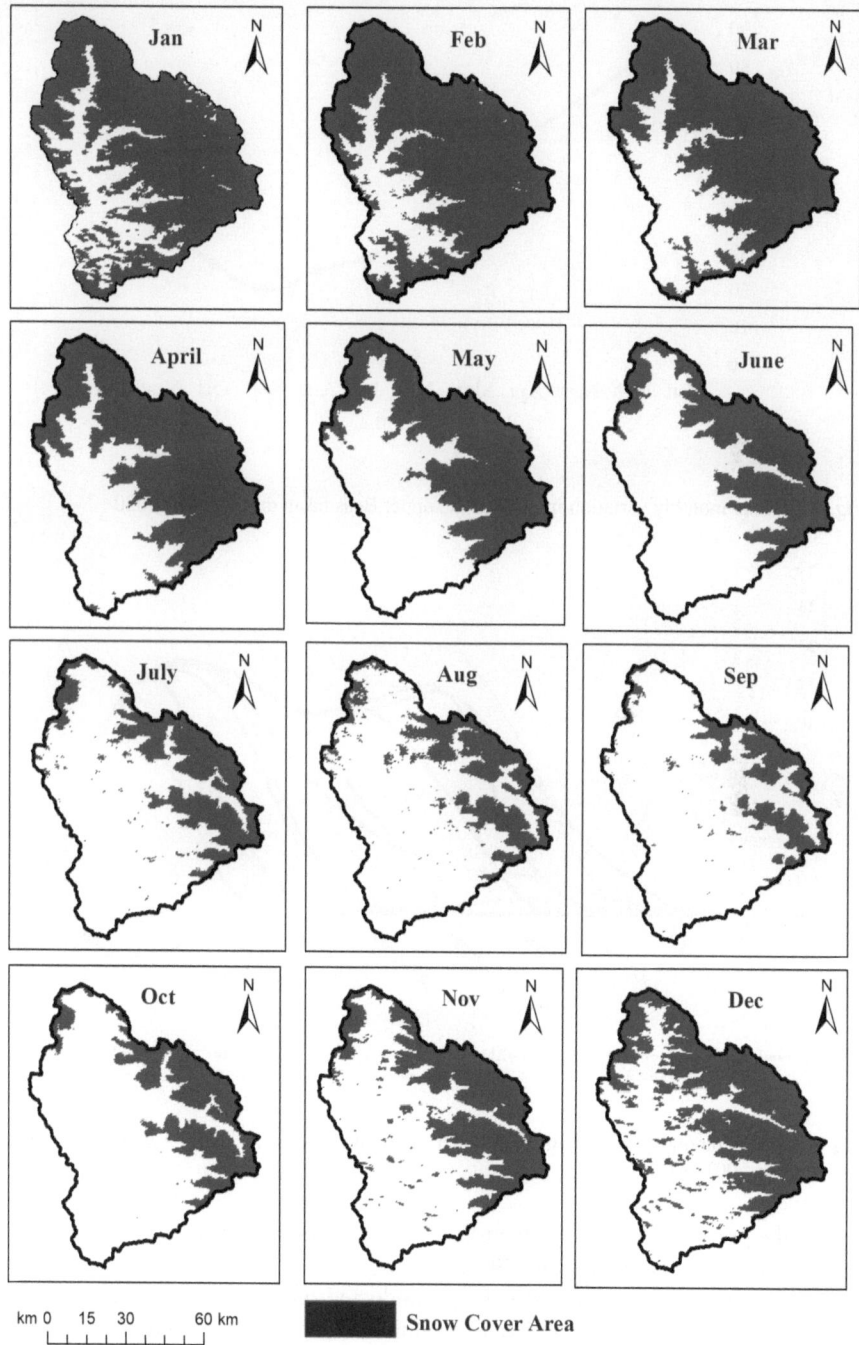

Fig. 4.6 Spatial distribution of mean monthly SCA of the Beas basin during 2001–2020

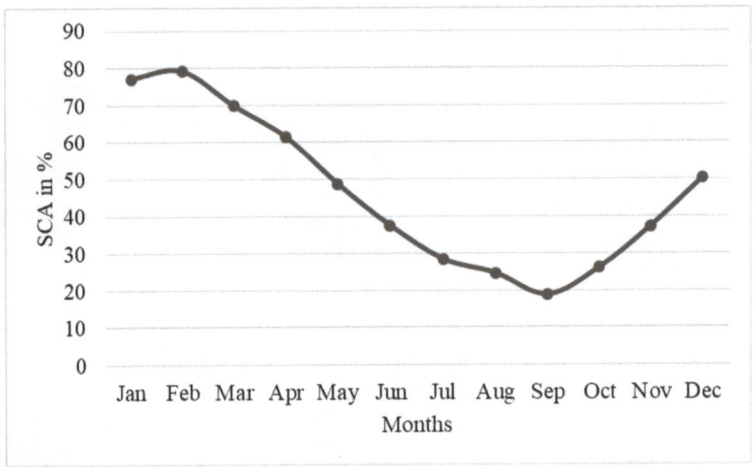

Fig. 4.7 Mean monthly variation in SCA of the upper Beas basin during 2001–2020

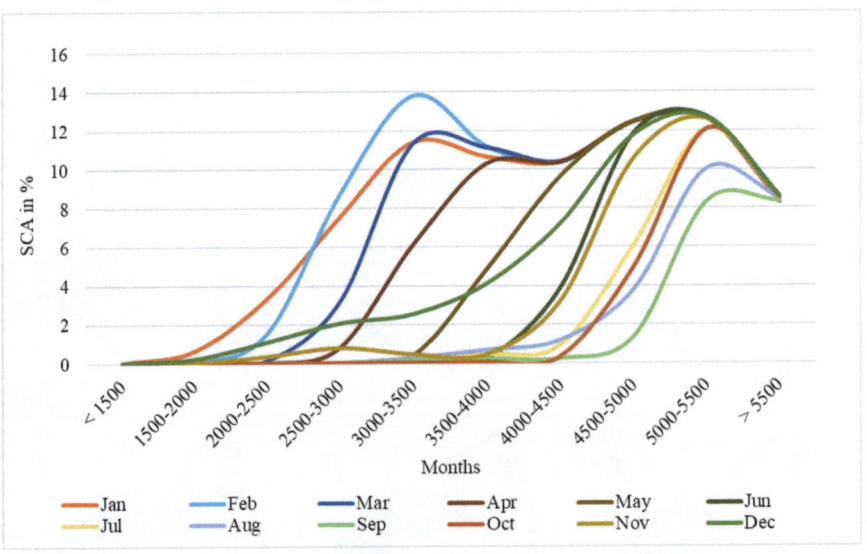

Fig. 4.8 Mean monthly variation in SCA under different elevation zones (m AMSL) of the Beas basin during 2001–2020

during the pre-monsoon season from 2000 to 2010, the research saw a strong detrimental impact of warming on the SCA.

A number of studies have also been conducted on the Indian Himalayas (IH) using remote sensing information to examine changes in the SCA (Rani 2018; Banerjee et al. 2021; Soni et al. 2021). They found evidence of a decline in SCA in the IH; however, the rate of loss varied over time and space. The SCA has been

Table 4.7 The trend in SCA
of the Beas River basin
during 2001–2020

Months	Trend (%/year)
Jan	−0.462
Feb	0.179
Mar	0.145
Apr	−0.301
May	0.114
Jun	0.379*
Jul	0.356
Aug	0.314
Sep	−0.027
Oct	−0.119
Nov	0.117
Dec	0.721

*Significant at the 0.05 level

significantly impacted by the warming trend in the upper Beas River basin between 2000 and 2010 (winter = 0.247%/year; pre-monsoon = 0.525%/year) (Rani and Sreekesh 2016). The pre-monsoon season sees a large decline in SCA, which causes the season's snow to melt earlier than usual (Rani 2014). In the Beas basin, Randhawa and Gautam (2019) found a pattern of increasing SCA in October by about 106.45%, with decreases of 16.18% in December (2018–2019) and 52.66% in each of November, February, and March. According to Sood et al. (2020), over the previous 10 years, the SCA variability over the IH, including the Karakoram Mountain ranges, has exhibited a shift of one month in snow accumulation and melting. According to Singh et al. (2021), SCA and total precipitation decreased by 3.2 km^2 and 64.7 cm, respectively, during the winter season (November–April) between 2003 and 2017. In the basin between 2003 and 2017, the findings revealed warming and decreasing precipitation. The whole of Karakoram Mountain, with the exception of the eastern Himalayas, has seen an increase in annual SCA (Dharpure et al. 2021). According to Shirsat et al. (2021), the basin's SCA will fall by between −10.1 (2050 RCP4.5) and −34.8%. (2090 RCP8.5). The baseline winter SCA of 98% of the catchment area decreases to 94% (RCP4.5) and in 2090, to 84.6% (RCP8.5). According to Bilal et al. (2019), the upper Indus basin (UIB) had a significant rise in SCA over 5000-m ASL between 2000 and 2017. However, in Asia's high mountains, where around 800 million people rely in part on glacial meltwater, melting glaciers also protect substantial populations from the consequences of drought stress (Pritchard 2019). Jabbar et al. (2020) found a very slight increase in SCA in the UIB from 1993 to 2019. According to Ali et al. (2020), the SCA dropped throughout the entire Indus basin and its sub-catchments from 2008 to 2018.

4.5   Summary

The findings show that all the LULC classes are increasing except cultivated land and barren land. The highest increase has been found in forest areas followed by water bodies. A slight rise has also been observed in the built-up area of the study area. This study has also assessed the change in settlement of the study area that shows that dense urban area is increasing more than other classes. The very-low-density rural settlements are declining in this study area. It shows that urbanization has increased in this study area during 1975–2015. This study has also understood the LULC in the upper Beas basin during 1980–2020 that shows that built-up area, cultivated land, and grassland are increasing in this study area. However, forest areas, barren land, and water bodies are declining in the study area. The trend in SCA shows that it is rising in all the months except January, April, September, and October during 2001–2020. The trend shows insignificant decline of SCA in pre-monsoon that can result in a rise of discharge. However, there is possibility of a decline in streamflow of the Beas river in the long run due to decline in SCA. Thus, it needs to be analyzed with fine-resolution data.

References

Ali S, Cheema MJ, Waqas MM, Waseem M, Awan UK, Khaliq T (2020) Changes in snow cover dynamics over the Indus Basin: evidences from 2008 to 2018 MODIS NDSI trends analysis. Remote Sens 12(17):2782

Anand J, Devak M, Gosain AK, Khosa R, Dhanya CT (2021) Spatio-temporal effect of climate and land-use change on water balance of the Ganga River basin. J Hydro-Environ Res 36:50–66

Awotwi A, Anornu GK, Quaye-Ballard JA, Annor T, Forkuo EK, Harris E, Agyekum J, Terlabie JL (2019) Water balance responses to land-use/land-cover changes in the Pra River Basin of Ghana, 1986–2025. Catena 182:104129

Banerjee A, Chen R, Meadows ME, Sengupta D, Pathak S, Xia Z, Mal S (2021) Tracking 21st century climate dynamics of the third pole: an analysis of topo-climate impacts on snow cover in the central Himalaya using Google Earth Engine. Int J Appl Earth Obs Geoinf 103:102490

Bilal H, Chamhuri S, Mokhtar MB, Kanniah KD (2019) Recent snow cover variation in the upper Indus basin of Gilgit Baltistan, Hindukush Karakoram Himalaya. J Mt Sci 16(2):296–308

Birhanu A, Masih I, van der Zaag P, Nyssen J, Cai X (2019) Impacts of land use and land cover changes on hydrology of the Gumara catchment, Ethiopia. Phys Chem Earth 112:165–174

Census of India (2011) Provisional population totals paper 1 of 2011: Himachal Pradesh. Available via   http://www.censusindia.gov.in/2011-prov-results/prov_data_products_himachal.html. Accessed 20 May 2015

Dharpure JK, Goswami A, Patel A, Kulkarni AV, Snehmani. (2021) Assessment of snow cover variability and its sensitivity to hydrometeorological factors in the Karakoram and Himalayan region. Hydrol Sci J 66(15):2198–2215

Gaur S, Bandyopadhyay A, Singh R (2021) Projecting land use growth and associated impacts on hydrological balance through scenario-based modelling in the Subarnarekha basin, India. Hydrol Sci J 66(14):1997–2010

Jabbar A, Othman AA, Merkel B, Hasan SE (2020) Change detection of glaciers and snow cover and temperature using remote sensing and GIS: a case study of the Upper Indus Basin, Pakistan. Remote Sens Appl Soc Environ 18:100308

Naha S, Rico-Ramirez MA, Rosolem R (2021) Quantifying the impacts of land cover change on hydrological responses in the Mahanadi River basin in India. Hydrol Earth Syst Sci 25(12):6339–6357

Prasena A, Shrestha DP (2013) Assessing the effects of land use change on runoff in Bedog Sub Watershed Yogyakarta. Indones J Geogr 45(1):48

Pritchard HD (2019) Asia's shrinking glaciers protect large populations from drought stress. Nature 569(7758):649–654

Randhawa SS, Gautam N (2019) Assessment of spatial distribution of seasonal snow cover during the year 2018–19 in Himachal Pradesh using space data. Available from: http://www.hpccc.gov.in/documents/SnowCoverAnalysis.pdf

Rani S (2014) Assessment of the influence of climate variability on the snow cover area of the Upper Beas River Basin. Unpublished M.Phil. Dissertation, CSRD, JNU

Rani S (2018) Evaluating snow cover changing trends of the Western Indian Himalaya. Spat Inf Res 26:103–112. https://doi.org/10.1007/s41324-017-0158-7

Rani S, Sreekesh S (2016) An analysis of pattern of changes in snow cover in the upper Beas River Basin, Western Himalaya. In: Raju N (ed) Geostatistical and geospatial approaches for the characterization of natural resources in the environment. Springer, Cham. https://doi.org/10.1007/978-3-319-18663-4_139

Shirsat TS, Kulkarni AV, Momblanch A, Randhawa SS, Holman IP (2021) Towards climate-adaptive development of small hydropower projects in Himalaya: a multi-model assessment in upper Beas basin. J Hydrol Reg 34:100797

Singh DK, Gusain HS, Dewali SK, Tiwari RK, Taloor AK (2021) Analysis of snow dynamics in Beas River basin, western Himalaya using combined terra–aqua MODIS improved snow product and in situ data during twenty-first century. In: Water, cryosphere, and climate change in the Himalayas. Springer, Cham, pp 115–128

Soni C, Chaudhary A, Sharma C (2021) Snow cover monitoring using topographical parameters for Beas river catchment area. In: Mapping, monitoring, and modeling land and water resources. CRC Press, Boca Raton, pp 297–310

Sood V, Singh S, Taloor AK, Prashar S, Kaur R (2020) Monitoring and mapping of snow cover variability using topographically derived NDSI model over north Indian Himalayas during the period 2008–19. Comput Geosci 8:100040

Woyessa YE, Welderufael WA (2021) Impact of land-use change on catchment water balance: a case study in the central region of South Africa. Geosci Lett 8(1):1

Chapter 5
Impact of Climate and LULC Changes on Hydrology

Abstract The climate and land use/land cover are the two most important variables impacting the hydrology of the basin. The SWAT hydrological model was built using data from climate, land cover, soil, and elevation from 1974 to 2020. Under all climate change scenarios, changes in monsoon-predicted mean streamflow in the year 2071 would range from –0.11% to 0.45%, while changes in winter-predicted mean streamflow would range from 0.28% to 1.23%. It might be ascribed to an increase in mean air temperature, which would result in early snowmelt and a change from snowfall to precipitation in the form of rain. The mean annual streamflow would range from 0.08% to 0.66% from the baseline by 2071 under all climate change scenarios. In the land cover scenarios, the reduction in monsoon streamflow by the end of the twenty-first century would be between –13% and –2.80%. By 2071, the change in streamflow during the pre-monsoon season in the basin would range from –5.37% to –23.78% due to fluctuations in the amount of snow cover in the basin. The percentage decrease in mean annual flow would vary from –5.88% to –2.41% in the basin. It is predicted that the impacts of changing land use and climate on streamflow would be more noticeable at the seasonal scale than the annual scale. Future research should examine the effects of shifting land cover and climate on the basin's hydrology.

Keywords Climate change · Land use/land cover · Scenarios · Hydrology · Future discharge

5.1 Introduction

The most significant factors affecting the basin's hydrology are the climate and land use/land cover (LULC). Changes in the climate and LULC are described in earlier chapters. The SWAT model is applied in the upper Beas basin up to Pandoh Dam due to the limited availability of discharge data. To estimate the upper Beas basin's hydrology's reaction to climatic factors and land cover change, future scenarios

S. Rani, *Climate, Land-Use Change and Hydrology of the Beas River Basin, Western Himalayas*, Advances in Asian Human-Environmental Research, https://doi.org/10.1007/978-3-031-29525-6_5

must be decided before running the SWAT hydrological model. Because the evaluation of climate variability and the analysis of LULC changes for the basin were completed up to 2020, the years 2020–2071 were chosen as the predicted period for the response of the basin hydrology under various scenarios of climatic and land cover change. The SWAT hydrological model was used in the current study to measure how these two parameters affected streamflow in the basin. This chapter discusses the future scenarios and their impact on basin hydrology by the late twenty-first century.

5.2 Future Scenarios

Depending on the use and development of the scenarios, there are several definitions of the word "scenario." According to the IPCC[1] (2007), scenarios are made to look at probable future emission routes, their underlying causes, and how they might be altered by policy changes. The current research adheres to the definition provided by Carter et al. (1994) as a credible, internally consistent, and coherent portrayal of a potential future condition of the world scenario. Results (impact on streamflow) are typically influenced by the sort of study scenarios used. For instance, if extreme climate and land cover change scenarios are chosen, extreme results may be expected, and vice versa. According to Feenstra et al. (1998), it is crucial to use different scenarios to demonstrate the uncertainty surrounding regional climate change since they give a comprehensive grasp of the variability in hydrological components. To aid policymakers in their decision-making, a variety of scenarios can be used to determine how sensitive a hydrological system is to changes in climate and land cover.

To understand how the future climate will affect streamflow, numerous kinds of research were conducted all around the world. General circulation models (GCMs) were the main source of climate change scenario data used in impact assessment studies (Faiz et al. 2018; Chen et al. 2019; Bekele et al. 2019; Sharma et al. 2022). In these investigations, the key influencing elements on the hydrological parameters are considered to be air temperature and rainfall. These studies have revealed both favorable and unfavorable effects of climate change on the hydrology of the basins by the middle and end of the twenty-first century, despite regional variability in their effects. Significant effects of land cover changes on streamflow have been documented in studies (Kumar et al. 2018; Garg et al. 2019; Chanapathi and Thatikonda 2020; Gaur et al. 2021). The type and spatial scale of transitions in land cover classes largely determine their impact. For instance, a decrease in forest cover can increase surface flow. The studies have either downscaled the regionally available anticipated land cover data or predicted future scenarios.

---

[1] https://www.ipcc.ch/publications_and_data/ar4/wg3/en/ch3s3-1-1.html

5.2.1 Climate Scenarios

In some research, streamflow changes under various future climate change scenarios were evaluated using the IPCC's Special Report on Emissions Scenarios (SRES) (Gebremeskel and Kebede 2018; Sharma and Babel 2018; Nasseri et al. 2019). Global and regional climate change scenarios are often scaled down to microlevel before the assessment is done since the projected global and regional climate data are of low resolution and cannot be used directly without being downscaled to micro size. Reliable and representative data are necessary for downscaling from the regional to the micro scale; otherwise, downscaling may be exceedingly challenging and perhaps inappropriate if the necessary observed climatic data are not available. For instance, the Himalayas do not have many weather observatories. Due to the harsh climate and high altitudes, there is also a considerable amount of uncertainty in the observed data. Therefore, rather than taking data from regional climate change scenarios, the study used a synthetic method to generate its climate change scenarios. Various researchers have developed climate change scenarios for impact assessment studies using the synthetic scenario method (Pervez and Henebry 2015; Schwank et al. 2014; Musau et al. 2015). This technique is based on the idea that "changes in air temperature and precipitation occur in modest increments," as stated by Feenstra et al. (1998). The approach is also said to be rapid, economical, and easy to construct, requiring generally few computer resources.

Future climate change scenarios were chosen for the current study based on assessments of the study area's climate variability and two reports (IPCC 2013; Krishnan and Sanjay 2017). According to the IPCC (2013), the majority of South Asia might see a rise in average annual temperatures of more than 2 °C by the middle of the twenty-first century compared to the average in the twentieth century. By the end of the twenty-first century, it may reach 3 °C and rise to more than 6 °C at high latitudes under a scenario with significant emissions (RCP 8.5). Average temperatures might climb by less than 2 °C in the twenty-first century, except in higher latitudes, when they may be up to 3 °C warmer (RCP 2.6). A high-emission scenario predicts that more rainfall will be quite probable at higher latitudes by the middle of the twenty-first century and across southern Asia by the late twenty-first century (RCP 8.5). By the mid-twenty-first century, more rain is predicted to fall at higher latitudes under the low-emission scenario (RCP 2.6), but significant changes in rainfall patterns are not likely at low latitudes. According to a report (Krishnan and Sanjay 2017), the warming in India over the short term (from 2016 to 2045) is roughly the same for all of the representative concentration pathways (RCP 2.6–8.5) scenarios, which fall between 1.08 and 1.44 °C. By 2045, the expected change in precipitation ranges from 0.16 to 0.27 mm.

By 2071, the basin's mean air temperature may rise by 2 °C (low-emission conditions) to 3 °C (high-emission conditions). The variation in rainfall has not been quantified in any of the reports. Therefore, the percentage changes in rainfall chosen for the current study are 10% and 15% by 2071. The study has chosen eight climate change (CC1–8) scenarios after taking into account the examination of air temperature,

Table 5.1 Climate change scenarios for the study area for 2020–2071

Scenarios	Air temperature (°C)	% change in rainfall
CC1	2	0
CC2	3	0
CC3	0	10
CC4	0	15
CC5	2	10
CC6	2	15
CC7	3	10
CC8	3	15

rainfall, and anticipated changes in the climatic variables listed in the reports above (Table 5.1). The baseline scenario period for the study, which reflects the current climatic conditions in the study area, is approximately 51 years (1969–2020). One aspect must be taken at a time and assumed to be constant to determine whether land or climate is the main influencing factor on the hydrology of the basin. Consequently, the study has changed the climate variable over time in scenarios of climate change while assuming the baseline scenario period's LULC conditions.

In CC2 and CC8 scenarios, air temperature and rainfall represent extreme climate change scenarios (Table 5.1). These are regarded as extreme climate change scenarios since the publications listed above describe such changes in air temperature and rainfall under high emissions scenarios (RCP 8.5), which are unlikely to occur in the upper Beas basin by the late twenty-first century. However, extreme climate change scenarios are useful for comprehending the extremely severe hydrological conditions that would soon exist in the study area. Additionally, it helps to comprehend the extent to which the basin's streamflow would be impacted by extreme climate conditions by the late twenty-first century. Each scenario was run for the same simulation period, except for the weather adjustment parameter in the SWAT model's modified climatic inputs (mean monthly air temperature and rainfall) (1969–2020). In reality, seasonal temperature patterns and subsequent distribution and frequency of precipitation events may be affected by changes in temperature and precipitation throughout the year. These factors aren't taken into account in the analysis for the simple reason that there isn't reliable information regarding how these changes are distributed. Additionally, it was believed that monthly variations in air temperature and precipitation would be consistent. To make the modelling more accurate for the years 2020–2071, monthly and seasonal changes in air temperature and rainfall were eliminated.

5.2.2   Land Cover Scenarios

Snow cover is therefore used as the base for creating land cover change scenarios because the decline in forest area won't be so significant that it would have an impact on streamflow (Rani and Sreekesh 2022). Another reason is that it is also supposed to be very sensitive to climate change and the basin receives roughly 35% of its flow from snow-melt, and snow cover is extremely sensitive to fluctuations in air temperature (Kumar et al. 2007). According to the analysis of SCA in the study area for the period 2000–2020, the highest and lowest SCA was found to be in February (80%) and September (20%). The mean SCA for these 2 months in the study area was calculated for the years 2001–2020 to determine land cover scenarios. To predict how SCA changes will affect the flow, three scenarios (LC1–3) were chosen based on descriptive data for SCA (Fig. 5.1).

5.3   SWAT Model Accuracy Assessment

5.3.1   Evaluation of Parameterization

The SWAT model needs to be calibrated to be regionalized and suitable for the Beas basin. In this study, elevation band, snow, and hydrology factors were chosen while taking the hydrological behavior of the Beas basin into consideration (Table 5.2). Eighteen different parameters altogether were chosen for the model calibration (Table 5.2). The SWAT model's elevation band-related parameters, including snow water content (SNOEB), temperature lapse rate (TLAPS), and precipitation lapse rate (PLAPS), were manually calibrated (Table 5.2). Because of the snow cover and elevational heterogeneity, the TLAPS and PLAPS were determined to be crucial for simulating the hydrological processes in the Beas basin.

Using the Thayyen and Dimri formula (2014), TLAPS and PLAPS were calculated based on the basin's mean air temperature. The optimal temperature lapse rate was -6 °C/km, which was shown to be consistent with other studies' use of temperature lapse rates between 5 and 7 °C/km (Thayyen et al. 2005; Baral et al. 2014). Although the precipitation elevation relationship is not necessarily linear (Immerzeel et al. 2014), there is a clear relationship between elevation and precipitation in the Himalayan region (Bookhagen and Burbank 2006). The upper Beas basin's valleys have an estimated PLAPS of -10 mm/km. According to Immerzeel et al. (2014), precipitation in the extremely high elevation parts of the Himalayas reduced as elevation increased. Due to a constraint in the SWAT model, which prevents the inclusion of TLAPS and PLAPS values by elevation bands and seasons, the equal temperature and precipitation lapse rate values were used for all height bands. As a result of snowfall in the Beas basin occurring above this elevation band, the initial snow water content (SNOEB) by elevation was taken to be zero below elevation band 3. Elevation bands 4, 5, 6, 7, and 8 each got SNOEB measurements of 30, 70, 80, 100, and 200 mm, respectively, due to the fact high-elevation bands accumulate more snow over time with less sublimation. Maity (2009) calculated SNOEB to be 30 mm near Bahang and Dhundi, where elevations range from 2900 to 3000 m (Table 5.2).

Fig. 5.1 Land cover
change scenarios of the
study area for the period
2020–2071

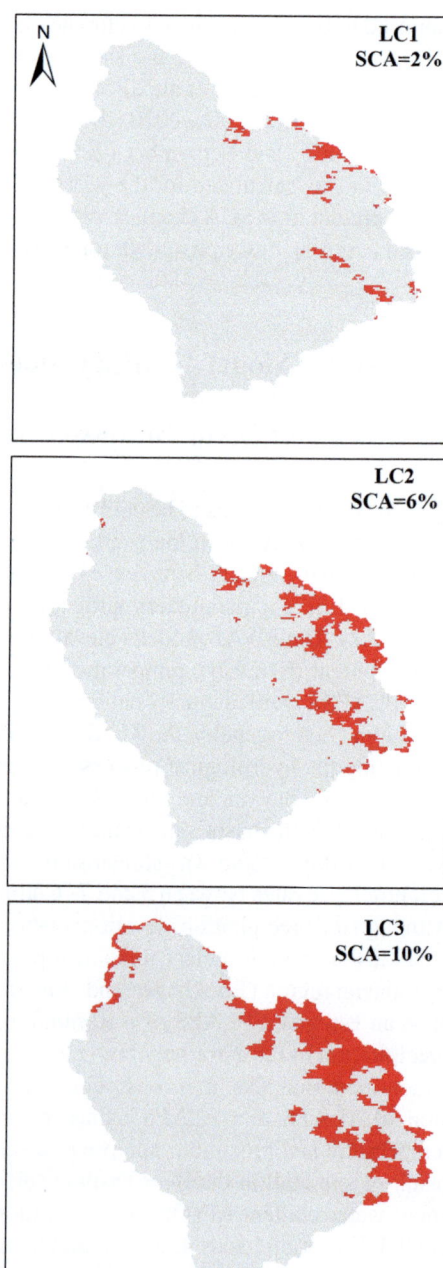

Table 5.2 SWAT model parameters and fitted values after calibration

Optimized parameters	Fitted value
Elevation band-related parameters	
Precipitation lapse rate (PLAPS) v	−10 mm/km
Temperature lapse rate (TLAPS) v	−6 °C/km
Snow water content (SNOEB) v	30–200 mm
Snow-related parameters-	
Rain/snow threshold (SFTMP) v	0 °C
Maximum melt coefficient (SMFMX) v	5 °C/mm/day
Minimum melt coefficient (SMFMN) v	2 °C/mm/day
Snowpack temperature lag factor (TIMP) v	0.68
Snowpack temperature melt factor (SMTMP) v	1
Areal snow coverage threshold CV100 (SNOCOVMX) v	0.68
Areal snow coverage threshold CV50 (SNO50COV) v	0.5
Hydrological parameters	
r__CN2.mgt	−0.15
v__GW_DELAY.gw (days)	31
v__SURLAG.bsn (days)	7
v__OV_N.hru	0.14
r__HRU_SLP.hru (m/m)	−1.8
v__GWQMIN.gw (mm)	3000
v__REVAPMN.gw (mm)	0
v__GW_REVAP.gw	0.2

Note: *r* relative change (%), *v* replace; CN2-SCS runoff curve number for moisture condition II; *GW_DELAY* groundwater delay, *SURLAG* surface runoff lag time, *OV_N* Manning's roughness for overland flow, *HRU_SLP* average slope steepness, *GWQMN* threshold depth of water in the shallow aquifer required for return flow to occur, *REVAPMN* threshold depth of water in the shallow aquifer for "revap" or percolation to deep aquifer to occur, *GW_REVAP* ground revap coefficient

A similar process was used to manually optimize the snow-pack temperature melt factor (SMTMP), maximum melt coefficient (SMFMX) on June 21 and minimum melt coefficient (SMFMN) on December 21, snowpack temperature lag factor (TIMP), areal snow coverage threshold CV100 (SNOCOVMX), and areal snow coverage threshold CV50 (SNO50COV) (Table 5.2). The rain/snow threshold (SFTMP) at Katrian is set at 0 °C based on observed snowfall and mean air temperature data from 1985 to 2015. It means that if the temperature in any sub-basin is less than 0 °C, the model treats precipitation as snowfall rather than rainfall. Because the basin received snowfall at this temperature, the snowpack temperature melt factor is set at 10 °C (Table 5.2). It means that snow melts in the basin occurs on days when the air temperature exceeds 1 °C. In the basin, the maximum and minimum melt coefficients were set at 5 °C/mm/day and 2 °C/mm/day, respectively. The snowpack temperature lag factor was set at 0.68, indicating that current-day air temperature has a greater influence on snowpack temperature and the previous day's snowpack temperature has a lesser influence (Table 5.2). The basin's snow coverage threshold CV100 and areal snow coverage threshold CV50 were set at 0.68 and 0.5 mm, respectively. The influence of the area depletion curve becomes more important in snowmelt processes as the value of CV100 increases (Neitsch et al. 2012).

The Latin hypercube one-factor-at-a-time (LH-OAT) approach of the SWAT model was used to assess the sensitivity of the parameter before executing the calibration (Abbaspour et al. 2015). This gives a general understanding of how to analyze both local and global sensitivity to assess the significance of a parameter. Sequential uncertainty fitting (SUFI2) was used to optimize eight important hydrological parameters in the SWAT-CUP (Table 5.2). The main representations of these factors were surface flow, groundwater, snow, evapotranspiration (ET), and the hydrological routing mechanism of the basin. These factors were selected based on Abbaspour's suggestions (2014). The following parameters were discovered to be frequently used in other studies to calibrate the model: SLSUBBSN (average slope length), GWQMN (threshold depth of water in the shallow aquifer required for return flow to occur), and GW REVAP (ground revap coefficient), CN2 (SCS flow curve number for moisture condition II based on different land cover and soil types), ESCO (soil evaporation compensation factor), and ALPHA BF (base flow alpha factor) (Shivhare et al. 2018; Tuo et al. 2018; Mengistu et al. 2019).

The automatic calibration technique SUFI2 was used to optimize the final fitted values while checking them for agreement with the basin characteristics and their underlying hydrological processes. With a base flow alpha factor value of 0.068, it was determined that the shallow aquifer basin had poor drainage and significant storage. GW of 31 days was shown to be the ideal amount of delay (Table 5.2). It implies that water will contribute to streamflow more gradually and enter the shallow aquifer more quickly. The low value of GWQMIN (3000 mm) helped to boost base flow, while the value of 0.2 for GW REVAP helped by reducing water transfer from the shallow aquifer to the root zone, which was necessary to imitate flow during low flow seasons. REVAPMN of the basin was configured to 0 mm (Table 5.2). It increases the amount of groundwater that is available to support streamflow. Due to a significantly higher groundwater table, enough soil moisture, and little transpiration, the modified EPCO (plant uptake compensation factor) value of 1 suggested that most of the water needed by plants would originate from the upper soil profile. Additionally, the ESCO value of 1 showed that additional water was taken from the upper level to reduce evaporative demand (Table 5.2). The surface flow lag coefficient, or SURLAG, was determined to be 7 (Table 5.2). More water is kept in reserve when the value of SURLAG decreases. It will delay surface runoff's release, smoothing the reach's predicted flow hydrograph.

5.3.2   Calibration and Validation

The model was run for the period 1974–2020, with the first 5 years being used as a warm-up period to set the hydrological conditions for the model. The SWAT model was calibrated and validated using the periods 1985–1993 and 1974–1985, respectively. Using monthly observed flow data from the Thalout station, the parameters described in the preceding section were used for model calibration at the basin level. To capture most of the measured data within the 95% prediction uncertainty, SUFI2 seeks to translate all uncertainties, such as model input, conceptualization, model parameters, and measured data, onto parameter ranges (Abbaspour et al. 2015; Narsimlu et al. 2015). The goal of SUFI2 is to collect most of the measured data within the smallest uncertainty band (Abbaspour 2014; Narsimlu et al. 2015). The P factor is the proportion of observed data that the 95PPU modelling solution encompasses. The 95PPU envelope's thickness is represented by the R factor (Abbaspour 2014). With a P-factor, the narrowest band, and an R-factor, the 95PPU band was intended to capture most of the measured data, including uncertainties. The additional performance indicators offered by SWAT-CUP, such as the coefficient of determination (R^2) and Nash-Sutcliffe (NS) (Nash and Sutcliffe 1970), were also included when evaluating the goodness of fit between the observation and the best simulation. The calibrated model was run for validation from 1974 to 1985, while the optimized parameters remained constant.

The average monthly observed and simulated streamflow in the Thalout basin over the calibration period (1985–1995) are shown in Fig. 5.2. After running the model 500 times with sensitive basin parameters, the graph, as shown in Fig. 5.2, represents the best simulation flow with a level of uncertainty band for the calibration period. In the areas of shaded gray, the simulation yielded 95% prediction uncertainty (95PPU). The amount of uncertainty in the basin's predicted flow is shown by the breadth of the darkened gray area. The P-factor was 0.67, meaning that the uncertainties could account for 67% of the reported daily streamflow, while the R-factor was 0.92. The P-factor and R-factor demonstrate the SWAT model's

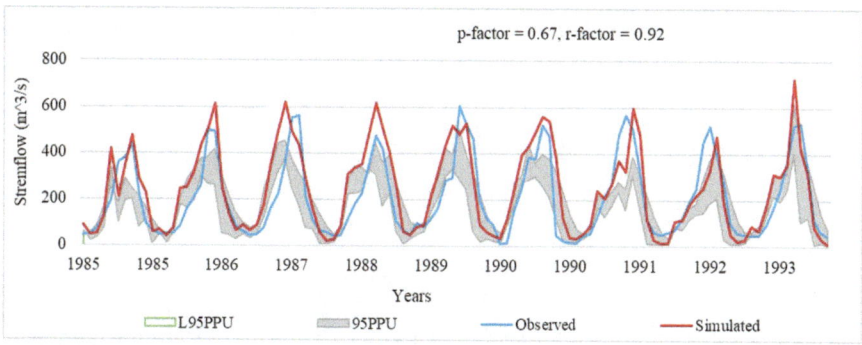

Fig. 5.2 Observed and simulated mean monthly flow at Thalout during 1985–1993 (blue color, observed flow; red color, best simulated flow)

dependability in modelling the streamflow in the Beas basin at Thalout station, in accordance with the findings. Graphical comparisons of the mean monthly stream-flow for the baseline (1974–2020), calibration (1985–1993), and validation (1974–1985) periods are shown in Fig. 5.3. Although some peak flow months were underpredicted during calibration, the model accurately tracked the observed mean monthly streamflow for the specified periods, and the underprediction was minimal during validation, perhaps as a result of the basin's low variability in precipitation.

It implies that the simulated and actual mean monthly flows correspond well across all three time periods. The relatively tiny difference between the mean monthly predicted flow and actual flow indicates reduced uncertainty (Table 5.3). The mean monthly observed flow exhibits greater fluctuation than the mean monthly predicted flow, as shown by the standard deviation (Table 5.3). For the calibration and validation periods, R^2 of the observed mean monthly flow with the simulated mean monthly flow was 0.81 and 0.81, respectively (Table 5.3). During the calibration and validation periods, the model predicted mean annual flow at Thalout by -1% and -4%, respectively (Table 5.3). The literature states that the SWAT model usually undervalues the major flow events since it is not designed to represent severe occurrences (Chu and Shirmohammadi 2004; Tolson and Shoemaker 2004; Cuo et al. 2013). Overall, the SWAT model was able to predict the hydrological conditions in the upper Beas basin with high accuracy.

The performance of the SWAT model was assessed for the baseline period, as shown in Fig. 5.4. The flow from February to May was overstated, whereas the flow from June to January was underestimated by the model (Fig. 5.4). From late March to May, the basin received mostly snowmelt runoff. The basin was fed by the monsoon from June to September. Seasonally, flow is overestimated during the winter and pre-monsoon seasons. The remaining seasons have underestimated flow during the baseline period. During the monsoon season, the difference between observed and simulated flow is greatest (Fig. 5.4).

During the monsoon season, the model is unable to match flow peaks. This may be because the rainfall data used for modelling is not always typical of the entire basin, which would indicate uncertainty. Another reason why the predicted flow is questionable is the absence of data on snowfall. As a result, the basin's snowfall data was not calibrated. At Thalout in the basin, there are relatively few annual observed

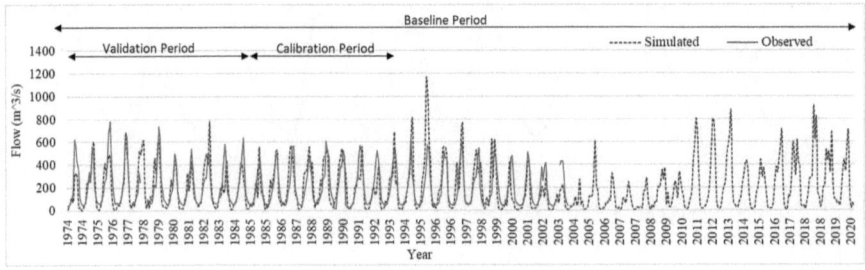

Fig. 5.3 Observed and simulated mean monthly flow at Thalout

Table 5.3 Statistics of mean monthly flow at Thalout during the calibration and validation periods

Type	Period	Time scale	Mean flow (m³/s) observed Mean	observed SD	Simulated Mean	Simulated SD	R²	NSE
Calibration	1985–1993	Monthly	208	170	205	175	0.81	0.63
Validation	1974–1985	Monthly	209	177	201	179	0.81	0.61

Note: *SD* standard deviation

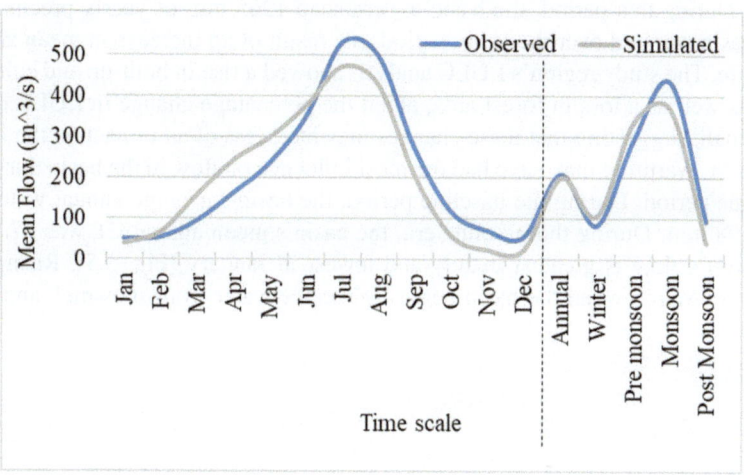

Fig. 5.4 Difference between observed and simulated mean monthly flow for the period 1974–2003

and simulated discharge changes. This suggests that the simulation is rather accurate. Reliable data is critical for improved SWAT model results (Fontaine et al. 2002). Another study discovered that the SWAT model performed better overall with a larger snowpack than with a smaller snowpack (Wang and Melesse 2005).

The SWAT model was used to simulate monthly, seasonal, and yearly streamflow although only monthly streamflow data by one station, Thalout, were used to calibrate and verify the model. This station is situated near the exit of the upper Beas basin. As a result, in subbasins where only calibrated parameters were applied based on the data from a single station, the uncertainty in the hydrological components would be larger. Due to errors or limits in the SWAT model as well as unknowns regarding future climatic conditions and emission scenarios, uncertainty may be introduced. The reanalysis data derived from a few weather stations (used in the model) which are mainly located in valleys along the lower course of the study area is not representative of the entire basin. The uncertainty in the simulated flow may have been increased by a lack of relevant climate data for higher elevations. The lack of available snow data for parameters such as snow melt rate, snow water equivalent, snow depth, and so on for the basin has also increased model uncertainty, particularly in spring season flow. Many of these uncertainties are difficult to quantify. As a result, interpreting the model results in the study necessitates taking these uncertainties into account.

5.4 Basin Hydrology During the Baseline Period (1969–2020)

Understanding the hydrological conditions of the basin is critical from 1969 to 2020. During this period, the basin experienced 1265 mm of yearly precipitation. The basin warmed over the study period as a result of an increase in mean air temperature. The study region's LULC analysis showed a rise in built-up and cultivated land as well as a loss in forest area; albeit the percentage change in LULC classes was small, suggesting that these changes may have less of an impact on the flow in the basin. Warming may have had the most influence on flow in the basin during the baseline period. During the baseline period, the basin's average annual water flow was 198 mm. During the baseline era, the basin's mean annual ET was 27.7 mm. The river's flow is greatest in July and lowest in January (Fig. 5.5). Rising limb begins in March when the basin begins to receive water from snowmelt and peaks

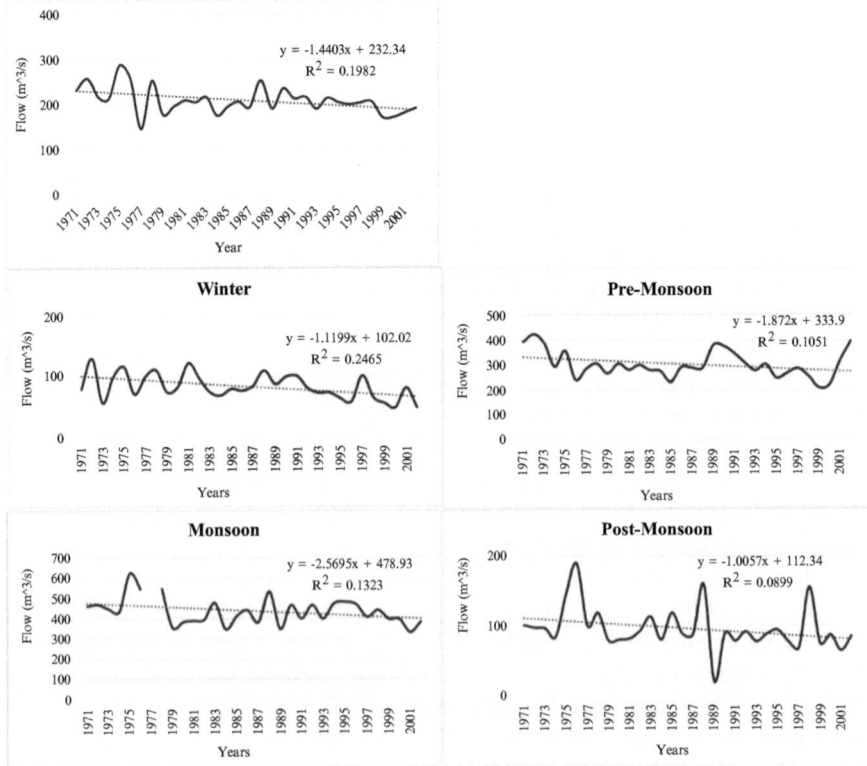

Fig. 5.5 Trend in mean annual and seasonal flow at Thalout in the study area

in July. Because the basin receives snowfall in December, the hydrograph curve gradually falls to its lowest point. Furthermore, snowmelt is nearly nonexistent throughout the winter.

For the analysis period, the trend in the mean annual flow at Thalout (Fig. 5.5) exhibits a declining tendency at a rate of 1.33 m³/s/year (Table 5.4). Even the mean flow declines throughout the year, from 1971 to 2002, a considerable downward trend was seen in the mean flow during the winter and post-monsoon seasons at rates of 1.26 m³/s/year and 0.72 m³/s/year, respectively (Table 5.4). Additionally, the mean flow during the analysis period showed a notable tendency to decline in January, February, March, July, November, and December (Table 5.4). In December, the mean monthly flow decreased at a rate of 0.47 m³/s/year, whereas in July, it decreased at a rate of 3.29 m³/s/year (Table 5.4).

A rise in air temperature would result in higher rates of plant transpiration and evaporation from water surfaces. This results in less outflow, although the basin also got more snowmelt water as the temperature increased. As a result, determining the precise reason for streamflow reduction during the baseline period would be impossible based on data from a single station. Furthermore, several studies have discovered a decrease in snowfall in the basin (Bhutiyani et al. 2010; Dimri and Dash 2012). As a result, this might have been a contributing factor to the basin's flow decline over time. Furthermore, when the temperature rises, the form of precipitation changes, resulting in changes in flow regimes. Other causes that might be causing the flow decline at Thalout must be investigated.

Table 5.4 Descriptive statistics and trend in mean flow at Thalout during 1971–2003

Time scale	Mann-Kendall test (Z)	Trend (m³/s/year)
Jan	−2.68**	−0.48**
Feb	−3.36**	−0.80**
Mar	−2.00*	−1.33*
Apr	−1.72	−2.33
May	−0.95	−1.39
Jun	−1.02	−1.85
Jul	−2.50*	−3.29*
Aug	−1.68	−3.33
Sep	0.00	−0.07
Oct	−1.45	−0.82
Nov	−3.35**	−0.70**
Dec	−2.78**	−0.47**
Winter	−2.71**	−1.26**
Pre-monsoon	−1.82	−2.00
Monsoon	−1.51	−1.90
Post-monsoon	−2.58**	−0.72**
Annual	−2.71**	−1.33**

Significant at 0.01** level and 0.05* level

5.5   Effect of Climate Change on Flow

The current study makes use of the SWAT hydrological model to examine potential changes in streamflow in the basin by the end of the century (2071) under several climate change scenarios. Although certainly, changes to land use would also occur in the basin in the near future, the current study assumed that it would be persistent to analyze the effect concerning the change in climatic variables, holding all other model components constant.

The baseline period is the SWAT model simulation for the period 1974–2020, as described in the preceding section. By assuming that no significant land use changes would occur by 2071 as a result of the current LULC analysis and forest policies, eight scenarios are taken into consideration to assess the number of changes in the streamflow of the upper Beas basin to temperature and rainfall. Since a 46-year time frame is preferred to represent baseline period conditions, each climate change scenario was run for the baseline period except those that used modified air temperature and rainfall inputs. In the upper Beas basin, the months of December to March are known as the "frozen period." During this time, the effects of glacier and snow melt on flow may be largely disregarded. About 7.2% of the total river flow is present during this time in the basin (Kumar et al. 2007). In the basin, snow melting starts in April and continues until October. An increase in air temperature would primarily affect the basin's ET rate, snow melt runoff, and precipitation types.

Figure 5.6 displays the percentage change in expected mean flow (by 2071) relative to the baseline (1974–2020) for various climate change scenarios. Pre-monsoon months would see a marginal increase of 0.14% and 0.41% in the expected mean flow in response to a 2 °C and 3 °C rise in mean air temperature, respectively (Fig. 5.6). The basin's susceptibility to the impact of a drop in snowpack level caused by a change in precipitation types and subsequent rise in snowmelt flow was shown by an increase in winter and pre-monsoon months' anticipated streamflow. Given the size of the glacier-covered area in the study area, the contribution of glaciers and snow melt to flow may be significant throughout the melting season.

Additionally, persistent increases in rainfall in the basin between 10% and 15% by 2071 would result in a linear rise in expected mean monthly streamflow relative to the baseline (Fig. 5.6). Pre- and post monsoon months would see the greatest rise in the predicted streamflow (Fig. 5.6). This also suggests that in contrast to air temperature, variations in rainfall had a more significant impact on flow in the study region. Winter and pre-monsoon months exhibit the highest increase in the expected mean flow by the 2071 period under CC5 and CC6 (Fig. 5.6). A rise in the mean air temperature would cause a switch from rain to snow in the type of precipitation, signifying less snowfall. The increase in mean air temperature in the basin would also cause early snow melting. Both of these elements would cause the estimated mean flow in the basin to increase throughout the winter.

Mean monthly flow would increase by the late century under CC7 and CC8 (2071). It demonstrates that compared to other circumstances, these (CC7–CC8) would have more pronounced flow alterations. This would result from a shift in

Time Scale	CC1	CC2	CC3	CC4	CC5	CC6	CC7	CC8	
Jan	0.77	1.22	0.30	0.45	1.08	1.23	1.52	1.68	
Feb	0.41	0.54	0.29	0.43	0.71	0.86	0.85	1.00	
Mar	0.44	0.69	0.27	0.41	0.72	0.87	0.97	1.12	
Apr	0.47	0.88	0.27	0.41	0.73	0.86	1.14	1.27	
May	0.22	0.39	0.35	0.51	0.48	0.62	0.66	0.79	
Jun	0.07	0.43	0.33	0.51	0.33	0.46	0.68	0.80	
Jul	-0.05	-0.10	0.30	0.45	0.23	0.38	0.29	0.48	
Aug	-0.16	-0.38	0.28	0.42	0.18	0.33	-0.08	0.07	
Sep	-0.16	0.01	0.32	0.48	0.31	0.56	0.34	0.50	
Oct	-0.41	-0.21	0.38	0.58	-0.04	0.15	0.21	0.42	% change in flow
Nov	-0.03	0.21	0.27	0.43	0.30	0.46	0.51	0.68	-1.5
Dec	0.73	1.09	0.33	0.50	1.07	1.24	1.42	1.59	-1.0
Annual	0.08	0.20	0.31	0.46	0.38	0.54	0.50	0.66	-0.5
Winter	0.49	0.81	0.28	0.42	0.77	0.91	1.09	1.23	0.0
Pre-Monsoon	0.14	0.41	0.34	0.51	0.40	0.53	0.67	0.79	0.5
Monsoon	-0.11	-0.19	0.30	0.45	0.23	0.39	0.15	0.32	1.0
Post-Monsoon	-0.28	-0.07	0.34	0.53	0.07	0.25	0.31	0.50	1.5

Fig. 5.6 Percentage change in predicted mean flow by 2071 with reference to the baseline period (1974–2020)

rainfall and an increase in mean air temperature from 2 °C to 3 °C (Fig. 5.6). A rise in mean air temperature would inevitably cause a change in precipitation type from rain to snow, which would result in a short-term decline in the snowpack. Early snowmelt runoff would occur in the basin at the same time. Overall, the estimated mean flow in the basin would increase. Under CC1 and CC2, the predicted mean annual flow in the basin would change by 0.08% to 0.20%, respectively (Fig. 5.6). Average yearly streamflow was projected to rise by 0.31% and 0.46% under CC3 and CC4, respectively (Fig. 5.6). The streamflow in the basin ranged between 0.38% and 0.66% in the other scenarios (Fig. 5.6). A rise in air temperature would typically result in an increase in ET, which would lead to a drop in soil water content, overall water output, and groundwater recharge. As a result of having less base flow, lateral flow, and surface flow, drier soil reduced water production. Although the study region has less ET than dryer places due to its highland environment, it obtains water from snowfall. Thus, a surge in surface flow would result from a change in air temperature, increasing the basin's water supply.

The results are consistent with the previous report (Lutz et al. 2016) and studies (Ragettli et al. 2016; Shukla et al. 2021). Nie et al. (2021) emphasized that such modifications might have a detrimental effect on the infrastructure and communities in the downstream areas, including hydropower and irrigated agriculture systems, by causing water flow to become more severe and unpredictable. Chandel and Ghosh (2021) found that the early part of the year will have higher flows, and the later portion of the year lower flows in the Satluj basin, Western Himalayas (RCP

4.5 and 8.5 scenarios). Grover et al. (2022) found an increase in discharge with a shift in the seasonal discharge pattern of the neighboring Chenab basin of the Western Himalayas by 2100 under RCP 4.5 and 8.5 scenarios. Momblanch et al. (2020) emphasize the need for mitigation and more aggressive adaptation measures to mitigate the effects of climate change in particularly climate-sensitive catchments. Considering the ongoing changes in the IH, Dimri et al. (2021) stated that a key component of successful climate change adaptation and response plans is more systematic and coordinated monitoring of the climate and its effects. Hassan et al. (2019) calculated streamflow for future climate change projections in the Northwestern upper Indus basin (UIB) using the RCP4.5 and RCP8.5 scenarios. Throughout the winter and spring seasons, significant streamflow changes were projected. Overall, because of the predicted increase in medium and high flow, the UIB should prepare for greater floods. The average annual runoff was found to be progressively increasing by the end of the twenty-first century. Hussain et al. (2020) evaluated the spatial and temporal changes of terrestrial water storage in the UIB for the study period of January 2003 to December 2016 and found negative trends of −4.35–0.38 mm/year. According to Dahri et al. (2021), the Indus-Tarbela inflows are more likely to increase than the inflows into the Kabul, Jhelum, and Chenab Rivers. Extreme climatic scenarios forecast a considerable increase in peak flow magnitudes and attainment 1 month earlier for all river gauges in the Indus basin. According to Nazeer et al. (2022), future glacier melt would increase by 30–265%, whereas UIB flow would increase by 23–126%. All heights will contribute more glacier melt than they did previously, according to the simulations, but the highest elevations will contribute far more. Studies were prepared by Azam et al. (2021) based on projections of rainfall, total runoff, and glacier/snow melt in the HK region. He asserts that the contributions of glaciers and snow melt, respectively, range from 2% to 50% and from 22% to 65% in the various basins of the HK region. Due to overuse in recent decades as a result of groundwater supplies not being enhanced to keep up with population development, groundwater levels in the Indus basin have decreased (Akhter et al. 2021). At UIB, the Baig et al. (2021) glaciers alone are responsible for two-thirds of the annual 28 river flows, while precipitation and snowfall each provide 19% and 11%. In UIB, the annual river flows will decrease overall by 16%, with considerable seasonal variations. According to Latif et al. (2021), the UIB witnessed a significant increase in summer flow from 1968 to 2013 and a reduction in August and September. Shah et al. (2020) found that the annual average precipitation will increase by 2.4–2.5% and 6.0–4.6% under RCP4.5 and RCP8.5 (mid- to late century), respectively. They also found an increase in the flow of 19.24% and 16.78% under RCP4.5, during the mid and late century, respectively. In the mid and end of the century, the flow will rise by 20.13% and 15.86%, respectively, according to RCP8.5. According to Wijngaard et al. (2017), by the end of the twenty-first century, the UIB, Ganges, and Brahmaputra River basins will probably grow due to future mean discharge and high flow conditions. Ashraf and Rehman (2019) found a modestly positive correlation between mean temperature, rainfall, and river discharges from the sub-basins in the UIB. Kiani et al. (2021) claim that under RCP4.5 and RCP8.5, the UIB's October and April estimated inflow increases are the greatest (37.99% and 65.11% at 1.5 °C and 2.0 °C, respectively). These hydrological changes are being brought on by an increase in the snow and glacier melt contribution, which is particularly pronounced at a warming level of 2.0 °C in the UIB. According to Ougahi et al. (2022), the average annual water yield from the glacial-nival regime increased, with river flow peaks happening 1 month earlier in the UIB. Summer precipitation was forecast to increase (RCP8.5: +36.7%) compared to the baseline (1974–2004), whereas winter precipitation was projected to decrease (RCP8.5: −16.9%). According to Khan et al. (2020), the UIB will see annual flow increases throughout the twenty-first century; curiously, these gains are higher in the middle years (2041–2070) than at the end of the century (2071–2100).

5.6 Effect of Land Cover Changes on Flow

In addition to climate, LULC is thought to be a significant factor influencing streamflow in an area. The baseline scenarios' time frame corresponds to the basin's LULC circumstances as of 2000. About 21% of the basin was covered with snow in 2000. Although the climate would change in the near future, the current study used the baseline climate when applying the land cover change scenarios to the SWAT model. Under scenario LC1, in the future (2071), the drop in mean monthly flow would range, in terms of relative change, from −1.29% to −42.05% (Fig. 5.7). Snowmelt and rainfall both provided water to the basin during the monsoon season. Although climatic variables in scenario LC1 are compared to baseline values, SCA is taken at roughly 2%, the lowest level across all land cover scenarios. The estimated mean streamflow in the basin by 2071 would be significantly reduced by a reduction in SCA from 26 (baseline period) to 2% (LC1) because snowmelt in the basin would be significantly reduced (Fig. 5.7). The increase in snow cover from 2 (LC1) to 6% (LC2) under scenario LC2 would cause a smaller reduction in the estimated mean streamflow than under the LC1 scenario (Fig. 5.7). By the end of the century, the basin's mean monthly streamflow would increase due to a slight increase in snowmelt runoff. Except for post-monsoon and winter, all scenarios showed a decrease in predicted mean streamflow (concerning the baseline period). Under every scenario, the annual discharge in the study area would decrease between 2.41% and 5.88%. All seasons, excluding the post-monsoon, have seen this state in the basin.

5.7 Summary

Climate, land cover, soil, and elevation data were used to put up the SWAT hydrological model for the period 1974–2020. To comprehend the uncertainty in the model's predicted flow data, an accurate evaluation of the basin model was also carried out. The model's performance was acceptable at the monthly and annual scales, but

Time Scale	LC1	LC2	LC3	
Jan	0.65	1.20	1.14	
Feb	0.33	0.33	0.30	
Mar	-1.29	-0.93	-0.81	
Apr	-7.01	-2.84	-2.45	
May	-19.68	-5.15	-4.70	
Jun	-27.87	-6.32	-5.93	
Jul	-42.05	-6.99	-6.48	
Aug	-16.98	-3.56	-3.16	
Sep	19.95	5.88	5.90	
Oct	17.03	20.08	19.46	% change in flow
Nov	4.77	11.56	11.06	-30.0
Dec	1.59	4.64	4.50	-20.0
Annual	-5.88	-2.73	-2.41	-10.0
Winter	-1.15	-1.24	-1.04	0.0
Pre-Monsoon	-23.78	-5.79	-5.37	10.0
Monsoon	-13.03	-3.17	-2.80	20.0
Post-Monsoon	10.90	17.23	16.65	30.0

Fig. 5.7 Percentage change in mean flow (by 2071) with reference to the baseline (1974–2020) in the study area

there were notable disparities between peak streamflows that was observed and simulated because of the study's sparse use of gauge stations and the potential underrepresentation of localized high-intensity rainfall. As a result, the use of data from a small number of meteorological stations may have led to an underestimating of severe occurrences and peak flows during the monsoon. The lack of snowfall data was a further source of uncertainty in the simulated mean flow.

The monthly, seasonal, and annual flow trends at Thalout in the basin between 1974 and 2020 show a clear downward trend. Though it can be challenging to pinpoint a precise reason for the decrease in streamflow at Thalout, research has shown that a warmer temperature has led to a loss in snowfall over time in the basin. Eight scenarios were developed with the assumption that significant land use changes won't occur anytime soon to assess the changes in the upper Beas basin's expected streamflow to temperature and rainfall (2071).

According to air temperature scenarios (CC1–CC2), the winter season has the largest increase in flow, which is followed by pre-monsoon by 2071. A change in precipitation from snowfall to rainfall or an increase in snow melting under warmer climate conditions might lead to an increase in winter flow. Additionally, these two processes have the potential to simultaneously boost streamflow throughout the basin's winter season. The spike in snowmelt runoff owing to the warmer environment may be the cause of the expected increase in flow during the pre-monsoon season. Given the strong need for hydropower and agriculture during this season, higher anticipated flows would be advantageous for the people.

Additionally, the model has projected a significant increase in flow during the pre-monsoon and post monsoon relative to baseline under the climate change scenarios (CC3 and CC4) that take into account a change in rainfall. It might be attributable to early snow melting.

In climate change scenarios (CC5–CC8) when both rainfall and air temperature change at the same time, a significant increase in flow is projected for winter, followed by pre-monsoon. It is a result of the change in precipitation type from snowfall to rainfall throughout the winter. Snowmelt rates rise due to the warmer temperature during these seasons.

In the future (2071), concerning the baseline, changes in monsoon-predicted mean streamflow would vary between −0.11% and 0.45%, whereas changes in winter predicted mean streamflow would vary between 0.28% and 1.23% for all climate change scenarios. It may be attributed to a rise in mean air temperature, which would cause early snowmelt and a transition from snowfall to rainfall in the form of precipitation. Under all climate change scenarios, the mean annual streamflow would change by 2071, varying from 0.08% to 0.66% from the baseline. By 2090, Shirsat et al. (2021) estimated that the upper Beas basin will see warming, increased precipitation, rising ET, and falling SCA. This research predicts that by 2090, the average annual streamflow would shift by −12% to 14.6% (RCP4.5) and by −18.6% to 48.1% (RCP8.5). The contributions of various hydrologic components alter as a result of changes in the catchment's cryospheric constituents (snowmelt, ice melt, rainfall-runoff, and baseflow). In all climate change scenarios, it is found that the contribution of snow and glacier melt runoff to total streamflow steadily declines, but the amount of rainy runoff rises toward the end of the century.

By the end of the twenty-first century, the percentage decrease in monsoon streamflow would range from −13% to −2.80% under the LULC scenarios. While variations in the amount of snow cover in the basin would cause the change in streamflow during pre-monsoon to range in the basin from −5.37% to −23.78% by 2071. The basin would have a range of −5.88% to −2.41% in the percentage change in mean annual flow. This is since during the monsoon, the flow receives water from both snowmelt and rain, but during the pre-monsoon, snowmelt is the major source of water for the flow. Snow cover reduction would certainly decrease the volume of snowmelt in the basin in the near future. There is a reduction in peak streamflow due to spring melting conditions for all three scenarios, but the time of expected peak streamflow is unaffected in the hydrograph for the baseline and predicted period. According to Haleem et al. (2022), from 2000 to 2013, climate change had a higher impact on river runoff in the UIB (61.61%) than land use change (38.39%). In the future, deeper runoff is anticipated as a result of both climate and land use changes. Climate change has a larger influence (12.76–25.92%) than land use change (0.37–1.1%).

The effects of climate and land use change on streamflow are anticipated to be more pronounced at the seasonal scale than the annual scale, possibly due to the expectation of low interannual variability in precipitation and the assumptions that historical spatial and temporal distributions of humidity, solar radiation, and wind speed will remain true for the foreseeable future. The basin's population's livelihood and hydropower production would benefit from increased water availability during the winter, pre-monsoon, and post monsoon seasons. However, there is also a need for more credible climatic data to help with better water management and policymaking by lowering the uncertainty in the forecast flow by the model. Overall, the study concluded that by 2071, changes in snow cover will be the dominant factor impacting streamflow variations in the basin. Future studies should include how changing climate and land cover together have an impact on the hydrology of the basin.

References

Abbaspour KC (2014) SWAT-CUP manual (SWAT calibration and uncertainty programs). Available via http://swat.tamu.edu/media/114860/usermanual_swatcup.pdf. Accessed 13 Jan 2014

Abbaspour KC, Rouholahnejad E, Vaghefi SR, Srinivasan R, Yang H, Kløve B (2015) A continental-scale hydrology and water quality model for Europe: calibration and uncertainty of a high-resolution large-scale SWAT model. J Hydrol 524:733–752

Akhter G, Ge Y, Iqbal N, Shang Y, Hasan M (2021) Appraisal of remote sensing technology for groundwater resource management perspective in Indus Basin. Sustainability 13(17):9686

Ashraf A, Rehman H (2019) Upstream and downstream response of water resource regimes to climate change in the Indus River basin. Arab J Geosci 12(16):1–0

Azam MF, Kargel JS, Shea JM, Nepal S, Haritashya UK, Srivastava S, Maussion F, Qazi N, Chevallier P, Dimri AP, Kulkarni AV (2021) Glaciohydrology of the Himalaya-Karakoram. Science 373(6557):eabf3668

Baig S, Sayama T, Yamada M (2021) Impacts of climate change on river flows in the Upper Indus Basin and its Subbasins. https://doi.org/10.21203/rs.3.rs-404691/v1

Baral P, Kayastha RB, Immerzeel WW, Pradhananga NS, Bhattarai BC, Shahi S, Galos S, Springer C, Joshi SP, Mool PK (2014) Preliminary results of mass-balance observations of Yala Glacier and analysis of temperature and precipitation gradients in Langtang Valley, Nepal. Ann Glaciol 55(66):9–14

Bekele D, Alamirew T, Kebede A, Zeleke GM, Melesse A (2019) Modeling climate change impact on the hydrology of Keleta watershed in the Awash River basin, Ethiopia. Environ Model Assess 24(1):95–107

Bhutiyani MR, Kale VS, Pawar NJ (2010) Climate change and the precipitation variations in the northwestern Himalaya: 1866–2006. Int J Climatol 30(4):535–548

Bookhagen B, Burbank DW (2006) Topography, relief, and TRMM-derived rainfall variations along the Himalaya. Geophys Res Lett 33(8):L08405

Carter TR, Parry ML, Harasawa H, Nishioka S (1994) IPCC technical guidelines for assessing climate change impacts and adaptations. Working Group II of the Intergovernmental Panel on Climate Change, University College, London. Available via http://www.climatechange.gov.bd/sites/default/files/IPCC_TechnicalGuidelinesAssessingCCImpacts.pdf. Accessed 20 Mar 2020

Chanapathi T, Thatikonda S (2020) Investigating the impact of climate and land-use land cover changes on hydrological predictions over the Krishna River basin under present and future scenarios. Sci Total Environ 721:137736

Chandel VS, Ghosh S (2021) Components of Himalayan river flows in a changing climate. Water Resour Res 57(2):e2020WR027589

Chen Y, Marek GW, Marek TH, Moorhead JE, Heflin KR, Brauer DK, Gowda PH, Srinivasan R (2019) Simulating the impacts of climate change on hydrology and crop production in the Northern High Plains of Texas using an improved SWAT model. Agric Water Manag 221:13–24

Chu TW, Shirmohammadi A (2004) Evaluation of the SWAT model's hydrology component in the piedmont physiographic region of Maryland. Trans ASAE 47(4):1057–1073. https://doi.org/10.13031/2013.16579

Cuo L, Zhang Y, Gao Y, Hao Z, Cairang L (2013) The impacts of climate change and land cover/ use transition on the hydrology in the upper Yellow River Basin, China. J Hydrol 502:37–52

Dahri ZH, Ludwig F, Moors E, Ahmad S, Ahmad B, Ahmad S, Riaz M, Kabat P (2021) Climate change and hydrological regime of the high-altitude Indus basin under extreme climate scenarios. Sci Total Environ 768:144467

Dimri AP, Dash SK (2012) Wintertime climatic trends in the western Himalayas. Clim Chang 111(3):775–800

Dimri AP, Allen S, Huggel C, Mal S, Ballesteros-Canovas JA, Rohrer M, Shukla A, Tiwari P, Maharana P, Bolch T, Thayyen RJ (2021) Climate change, cryosphere and impacts in the Indian Himalayan Region. Curr Sci 120:774

Faiz MA, Liu D, Fu Q, Li M, Baig F, Tahir AA, Khan MI, Li T, Cui S (2018) Performance evaluation of hydrological models using ensemble of general circulation models in the northeastern China. J Hydrol 565:599–613

Feenstra JF, Burton I, Smith JB, Tol RSJ (1998) Handbook on methods for climate change impact assessment and adaptation strategies. Available via http://lib.icimod.org/record/13767/files/7157.pdf. Accessed 24 Aug 2020

Fontaine TA, Cruickshank TS, Arnold JG, Hotchkiss RH (2002) Development of a snowfall–snowmelt routine for mountainous terrain for the soil water assessment tool (SWAT). J Hydrol 262(1–4):209–223

Garg V, Nikam BR, Thakur PK, Aggarwal SP, Gupta PK, Srivastav SK (2019) Human-induced land use land cover change and its impact on hydrology. HydroResearch 1:48–56

Gaur S, Bandyopadhyay A, Singh R (2021) Projecting land use growth and associated impacts on hydrological balance through scenario-based modelling in the Subarnarekha basin, India. Hydrol Sci J 66(14):1997–2010

Gebremeskel G, Kebede A (2018) Estimating the effect of climate change on water resources: integrated use of climate and hydrological models in the Werii watershed of the Tekeze river basin, Northern Ethiopia. Agric Nat Resour 52(2):195–207

Grover S, Tayal S, Sharma R, Beldring S (2022) Effect of changes in climate variables on hydrological regime of Chenab basin, western Himalaya. J Water Clim Chang 13(1):357–371

Haleem K, Khan AU, Ahmad S, Khan M, Khan FA, Khan W, Khan J (2022) Hydrological impacts of climate and land-use change on flow regime variations in upper Indus basin. J Water Clim Chang 13(2):758–770

Hassan M, Du P, Mahmood R, Jia S, Iqbal W (2019) Streamflow response to projected climate changes in the Northwestern Upper Indus Basin based on regional climate model (RegCM4. 3) simulation. J Hydro-Environ 27:32–49

Hussain D, Kao HC, Khan AA, Lan WH, Imani M, Lee CM, Kuo CY (2020) Spatial and temporal variations of terrestrial water storage in upper Indus basin using GRACE and altimetry data. IEEE Access 8:65327–65339

Immerzeel WW, Petersen L, Ragettli S, Pellicciotti F (2014) The importance of observed gradients of air temperature and precipitation for modeling runoff from a glacierized watershed in the Nepalese Himalayas. Water Resour Res 50(3):2212–2226

IPCC (2007) Climate change 2007: impacts, adaptation and vulnerability. Contribution of working group II to the fourth assessment report of the intergovernmental panel on climate change, M.L. Parry, O.F. Canziani, J.P. Palutik of, P.J. van der Linden and C.E. Hanson, Eds., Cambridge University Press, Cambridge

IPCC (2013) Climate change 2013: impacts, adaptation, and vulnerability. Contribution of working group II to the fifth assessment report of the intergovernmental panel on climate change. Cambridge University Press, Cambridge

Khan AJ, Koch M, Tahir AA (2020) Impacts of climate change on the water availability, seasonality and extremes in the Upper Indus Basin (UIB). Sustainability 12(4):1283

Kiani RS, Ali S, Ashfaq M, Khan F, Muhammad S, Reboita MS, Farooqi A (2021) Hydrological projections over the Upper Indus Basin at 1.5° C and 2.0 °C temperature increase. Sci Total Environ 788:147759

Krishnan R, Sanjay J (2017) Climate change over India: an interim report. Centre for Climate Change Research Available via http://cccrtropmetresin/home/old_portalsjsp. Accessed 20 Oct 2017

Kumar V, Singh P, Singh V (2007) Snow and glacier melt contribution in the Beas River at Pandoh dam, Himachal Pradesh, India. Hydrol Sci J 52(2):376–388

Kumar N, Singh SK, Singh VG, Dzwairo B (2018) Investigation of impacts of land use/land cover change on water availability of Tons River Basin, Madhya Pradesh, India. Model Earth Syst Environ 4(1):295–310

Latif Y, Ma Y, Ma W (2021) Climatic trends variability and concerning flow regime of Upper Indus Basin, Jehlum, and Kabul river basins Pakistan. Theor Appl Climatol 144(1):447–468

Lutz A, Immerzeel WW, Bajracharya SR, Litt M, Shrestha AB (2016) Impacts of climate change on the cryosphere, hydrological regimes and glacial lakes of the Hindu Kush Himalayas: a review of current knowledge. International Centre for Integrated Mountain Development (ICIMOD). Available via https://lib.icimod.org/record/32320/files/icimodCCCRR3_016. pdf?type=primary

Maity DK (2009) Hydrological and 1 D hydrodynamic modelling in Manali Sub-Basin of Beas River, Himachal Pradesh, India. Unpublished M.Sc. Dissertation. IIRS, Dehradun. Available via www.iirs.gov.in/iirs/sites/default/files/StudentThesis/dilip_kumar.pdf. Accessed 20 May 2014

Mengistu AG, van Rensburg LD, Woyessa YE (2019) Techniques for calibration and validation of SWAT model in data scarce arid and semi-arid catchments in South Africa. J Hydrol Reg Stud 25:100621

Momblanch A, Beevers L, Srinivasalu P, Kulkarni A, Holman IP (2020) Enhancing production and flow of freshwater ecosystem services in a managed Himalayan river system under uncertain future climate. Clim Chang 162(2):343–361

Musau J, Sang J, Gathenya J, Luedeling E (2015) Hydrological responses to climate change in Mt. Elgon watersheds. J Hydrol Reg Stud 3:233–246

Narsimlu B, Gosain AK, Chahar BR, Singh SK, Srivastava PK (2015) SWAT model calibration and uncertainty analysis for streamflow prediction in the Kunwari River Basin, India, using sequential uncertainty fitting. Environ Process 2(1):79–95

Nash J, Sutcliffe J (1970) River flow forecasting through conceptual models: part I. A discussion of principles. J Hydrol 10(3):282–290

Nasseri M, Zahraie B, Tootchi A (2019) Spatial scale resolution of prognostic hydrological models: simulation performance and application in climate change impact assessment. Water Resour Manag 33(1):189–205

Nazeer A, Maskey S, Skaugen T, McClain ME (2022) Changes in the hydro-climatic regime of the Hunza Basin in the Upper Indus under CMIP6 climate change projections. Sci Rep 12(1):1–6

Neitsch S, Arnold J, Kiniry J, Williams J (2012) Soil and water assessment tool-theoretical documentation version 2009. Texas Water resources Institute, Texas. Available via http://swattamu-edu/media/99192/swat2009-theorypdf. Accessed 20 May 2020

Nie Y, Pritchard HD, Liu Q, Hennig T, Wang W, Wang X, Liu S, Nepal S, Samyn D, Hewitt K, Chen X (2021) Glacial change and hydrological implications in the Himalaya and Karakoram. Nat Rev Earth Environ 2(2):91–106

Ougahi JH, Cutler ME, Cook SJ (2022) Modelling climate change impact on water resources of the Upper Indus Basin. J Water Clim Chang 13(2):482–504

Pervez MS, Henebry GM (2015) Assessing the impacts of climate and land use and land cover change on the freshwater availability in the Brahmaputra River basin. J Hydrol Reg Stud 3:285–311

Ragettli S, Immerzeel WW, Pellicciotti F (2016) Contrasting climate change impact on river flows from high-altitude catchments in the Himalayan and Andes Mountains. Proc Natl Acad Sci 113(33):9222–9227

Rani S, Sreekesh S (2022) Assessment and prediction of land use/land cover changes of Beas basin using a modeling approach. In: Mountain landscapes in transition. Springer, Cham, pp 471–487

Schwank J, Escobar R, Girón GH, Morán-Tejeda E (2014) Modeling of the Mendoza river watershed as a tool to study climate change impacts on water availability. Environ Sci Pol 43:91–97

Shah MI, Khan A, Akbar TA, Hassan QK, Khan AJ, Dewan A (2020) Predicting hydrologic responses to climate changes in highly glacierized and mountainous region Upper Indus Basin. R Soc Open Sci 7(8):191957

Sharma D, Babel MS (2018) Assessing hydrological impacts of climate change using bias-corrected downscaled precipitation in Mae Klong basin of Thailand. Meteorol Appl 25(3):384–393

Sharma SK, Sinha RK, Eldho TI (2022) Hydrological impact assessment of climate change on a Tropical River Basin in Southern India. Copernicus Meetings

Shirsat TS, Kulkarni AV, Momblanch A, Randhawa SS, Holman IP (2021) Towards climate-adaptive development of small hydropower projects in Himalaya: a multi-model assessment in upper Beas basin. J Hydrol Reg 34:100797

Shivhare N, Dikshit PK, Dwivedi SB (2018) A comparison of SWAT model calibration techniques for hydrological modeling in the Ganga river watershed. Engineering 4(5):643–652

Shukla S, Jain SK, Kansal ML (2021) Hydrological modelling of a snow/glacier-fed western Himalayan basin to simulate the current and future streamflows under changing climate scenarios. Sci Total Environ 795:148871

Thayyen RJ, Dimri AP (2014) Factors controlling Slope Environmental Lapse Rate (SELR) of temperature in the monsoon and cold-arid glacio-hydrological regimes of the Himalaya. Cryosphere Discuss 8(6):5645–5686

Thayyen RJ, Gergan JT, Dobhal DP (2005) Slope lapse rates of temperature in Din Gad (Dokriani glacier) catchment, Garhwal Himalaya, India. Bull Glaciol Res 22:31–37

Tolson BA, Shoemaker CA (2004) Watershed modeling of the Cannonsville Basin using SWAT2000: model. Cornell University, Ithaca. Available via. https://ecommons.cornell.edu/bitstream/handle/1813/2710/2004-2.pdf;sequence=1. Accessed 30 May 2014

Tuo Y, Marcolini G, Disse M, Chiogna G (2018) A multi-objective approach to improve SWAT model calibration in alpine catchments. J Hydrol 559:347–360

Wang X, Melesse AM (2005) Evaluation of the SWAT model's snowmelt hydrology in a northwestern Minnesota watershed. Trans ASAE 48(4):1359–1376

Wijngaard RR, Lutz AF, Nepal S, Khanal S, Pradhananga S, Shrestha AB, Immerzeel WW (2017) Future changes in hydro-climatic extremes in the Upper Indus, Ganges, and Brahmaputra River basins. PLoS One 12(12):e0190224

Chapter 6
Initiatives on Climate Change Mitigation

Abstract Eight national missions were established by the National Action Plan for Climate Change (NAPCC) to implement "multi-pronged, long-term integrated solutions" in the context of climate change. To address these escalating climate-related issues, the Ministry of Environment, Forest, and Climate Change (MoEFCC) works with many ministries involved in power, renewable energy, urban development, research and technology, water resources, and agriculture. The Indian government is developing and implementing several measures to boost the nation's ability to produce solar energy. The Indian government also sponsors the Atal, Swachh Bharat, and Smart Cities for Sustainable Habitat missions. The Energy and Resources Institute (TERI) wrote reports on the missions' successful completion. Several initiatives, like the National Mission on Sustainable Habitat (NMSH) and National Mission for Sustaining the Himalayan Ecosystem (NMSHE), have a reduced greenhouse gas emission target in mind to ensure the long-term sustainability of the Indian Himalayas. Punjab and Himachal Pradesh are also helping to improve conditions in the basin through the State Action Plan on Climate Change (SAPCC). The development of highly effective solutions and oversight of their execution is essential for adapting to and mitigating climate change.

Keywords Mitigation · National Action Plan for Climate Change · Sustainability · Management · Implementation

6.1 Introduction

As we have seen in the previous chapters, climate and land use/land cover (LULC) changes are a matter of big concern in the basin. According to Edenhofer et al. (2012), humans are currently confronted with three interconnected challenges: eliminating global poverty, enabling development, and averting climate change. Despite India's collective efforts at the national and international levels to address the climate crisis, its consequences remain important challenges for India. Besides,

© The Author(s), under exclusive license to Springer Nature
Switzerland AG 2023
S. Rani, *Climate, Land-Use Change and Hydrology of the Beas River Basin,
Western Himalayas*, Advances in Asian Human-Environmental Research,
https://doi.org/10.1007/978-3-031-29525-6_6

India has a severe ecological, social, and economic challenge to prevent the fast loss of natural resources (Lolaksha and Anand 2017). Climate change mitigation is concerned with reducing and adapting while developing efficient means of managing these actions and their implications across several areas (Knieling and Filho 2013). The Kyoto Protocol (KP) and the United Nations Framework Convention on Climate Change (UNFCCC) are the key organizations for today's global climate governance from a legal perspective (United Nations 1992; Bernauer and Schaffer 2010). India's existing climate change policies and practices place a strong emphasis on choosing actions and completing missions. The present climate change strategy has seen significant modifications in terms of its visible initiatives at the national and international levels (Saryal 2018). Given the importance of rising challenges in the basin, this chapter attempts to synthesize climate and LULC initiatives at the international, national and regional levels.

6.2   Concern on Governance

Taking into account the emerging climate-related changes and their consequences, climate change governance deliberates on describing and clarifying climate change policies. It demands a coordinated response at all levels (Ostrom 2010). It necessitates governments taking an active role in bringing about adjustments in interest perceptions to sustain stable social majorities in support of implementing an active mitigation and adaptation policy regime (Meadowcroft 2009). The institutionalism method of monitoring systems that combines public and private, hierarchical, and network forms of active management has governance as a distinguishing element (Renate 2004). A rising industry, climate change governance is closely linked to state administrative systems; private player attitudes, including business interests; nongovernmental groups; and civil society, including both adaptation and mitigation measures (Knieling and Filho 2013). The conversation about environmental protection started in the 1950s (Wilde 2008), but the conversation about climate change started around 20 years later. Concern over climate security has increased as a result of certain significant occurrences in the twenty-first century (Stern and Stern 2017). Climate change has been identified as a key danger to the preservation of global peace and security by the United Nations Security Council (UNSC). Since its creation, the Intergovernmental Panel on Climate Change (IPCC) has disseminated climate change knowledge (Intergovernmental Panel on Climate Change (IPCC) undated). The Human Development Report (HDR) for 2007–2008 unequivocally supports the difficult mission of combating climate change (Sahu 2019).

The IPCC was created in 1988 by the World Meteorological Organization (WMO) and the United Nations Environment Programme (UNEP) to offer policymakers regular scientific assessments of climate change, its effects, and potential future dangers, as well as to suggest adaptation and mitigation measures. Its purpose was to assess relevant data from the fields of science, technology, and sociology to conceptualize the scientific underpinnings of the threat posed by anthropogenic climate change, as well as the desires for change and mitigation, on a wide, open, and transparent basis. Although the IPCC was a joint effort between the UNEP and WMO, it covered all subjects that were outside of their purview (Drexhage 2008). At the moment, India is most vulnerable to climate change. It has genuine aspirations for reaching a substantial outcome, as well as a growing awareness of its potential role in accomplishing such goals. Recently, there has been a shift in the Indian approach toward UNFCCC discussions and more imaginative climate policy action at the national and regional level. This trend toward "multilevel governance," (e.g., Ministry of Environment, Forest, and Climate Change (MoEFCC); Science and Technology; Ministries of New and Renewable Energy (MNRE); Urban Development; Water Resources; and Agriculture) with a more autonomous subnational character, affected policy at each level (Atteridge et al. 2012).

6.3   Global Climate Change Initiatives

In recent decades, international politics and diplomacy have increasingly focused on the subject of climate change. The "UN General Assembly" announced the "UN Conference on the Human Environment in Stockholm" in 1972 with the major objective of the conference being to organize plans for measures to be done by states and transnational organizations to protect and enhance the environment for mankind (Brisman 2011). The Brundtland Commission Report of 1987 used the term "sustainable development" (Adams and Schuurman 1993; Carter 2001). It was promoted during the 1992 Rio Earth Summit and subsequently approved by the UNFCCC. The IPCC (2014) provided scientific studies with complete technicalities and socioeconomic recommendations to the international community, primarily to UNFCCC parties, through periodic reviews (Saryal 2018). Initially, the Indian policy framework appeared to be widely philosophical and intellectual, asserting essentially that India is a developing country and that solving climate change was the duty of developed countries (Messner 2017; Saryal 2018). As a result, India's present climate change policy looks to be characterized by realism. A comprehensive revamp of India's present climate change strategy has begun, intending to foster robust interactions at both the national and international levels (Saryal 2018). International initiatives taken to combat the climate change impacts are listed in Table 6.1.

6.4   National Climate Change Initiatives

The Indian government has launched many measures both domestically and internationally to tackle climate change. At first, efforts were made to conserve the nation's natural resources, with less emphasis placed on combating climate change. Examples of these efforts include the creation of an environmental legal framework,

Table 6.1 Climate change mitigation initiatives at international level

Year	Initiative	Remarks
1972	Stockholm Declaration	The necessity of a common vision and set of principles to motivate and direct people throughout the world in preserving and improving the human environment
1980	The World Climate Research Programme (WCRP) was established by the World Meteorological Organization (WMO) in Geneva and the International Council of Scientific Unions (ICSU) in Paris	Climate research has received a significant boost, notably in the numerical simulation of atmospheric and oceanic processes
1987	The Brundtland Commission Report	"Our Common Future" acquainted with the idea of sustainable development coined the concept of sustainable development and described how it could be achieved
1988	IPCC establishment	By the UNEP and WMO To write and publish studies that provide readers a precise and current overview of the status of climate change science today
1989	United Nations General Assembly resolutions	The "Framework Convention" has been the subject of negotiations
1990	IPCC First Report "Intergovernmental Negotiating Committee (INC)" convened for UNFCC negotiations	The First IPCC Assessment Report (FAR) emphasized the significance of climate change as a problem having global implications and demanding cooperation on a global scale
1992	Rio "Earth Summit" The Convention on Biological Diversity (CBD)	In order to provide a broad agenda and new framework for international action on environmental and development issues that would guide international cooperation and development strategy in the twenty-first century, the UNFCCC was signed in Rio de Janeiro CBD is an international legal document that has been approved by 196 countries for "the protection of biological diversity, the sustainable use of its components, and the fair and equitable sharing of the benefits emerging from the usage of genetic resources"
1994	UNFCCC entered into effect	To counteract "dangerous human influence with the climate system," in part by stabilizing atmospheric greenhouse gas concentrations
1995	IPCC Second Assessment Report COP 1	Before the Kyoto Protocol was adopted in 1997, the IPCC's Second Assessment Report (SAR) gave crucial information for nations to consider Calls for governments to set clear, legally binding objectives and deadlines for decreasing greenhouse gas emissions in industrialized countries

(continued)

Table 6.1 (continued)

Year	Initiative	Remarks
1997	COP 3	KP agreed to cut emissions of six greenhouse gases by 5.2% compared to 1990 levels at some time between 2008 and 2012 at the COP 3 summit.
2001	IPCC Third Assessment Report COP 7	The implications of climate change and the need for adaptation were the main topics of the IPCC's Third Assessment Report (TAR) Marrakesh Accords created a new, more efficient, and legally binding method of resolving disputes
2002	COP 8	Key topics include vulnerability and adaptation, CDM, LULUCF, and mitigation, and there is a distinct South Asian focus
2005	COP 11	At COP 11, the Ad-Hoc Working Group on KP (AWG-KP) was established to examine "second commitment period" objectives. "Dialogue" on the long-term cooperative action (LCA) idea to establish their own carbon exchange was started at COP 11
2007	G8+5 Summit IPCC Fourth Assessment Report COP 13	G8+5 Summit: The committee agreed that man-made climate change was undeniably real and that there should be a worldwide system of emission caps and carbon emissions trading that applied to both developed and developing countries A post-Kyoto agreement, with an emphasis on keeping warming to 2 °C, was spelled out in IPCC's Fourth Assessment Report (AR4) from 2007 Shared vision, mitigation, adaptation, technology, and funding are the five core divisions of the Bali Action Plan. A long-term objective for emission reductions is part of the shared vision, which is a long-term plan for addressing climate change
2008	COP 14	The Conference approved the Delhi Declaration, in which parties pledged support for a number of topics, including women and health, ecosystem restoration, and combating climate change, including the commercial sector, the Peace Forest Initiative, and recovering 26 million hectares of India's damaged land
2009	Major Economies Forum on Energy and Climate (MEFEC) COP 15	Indicated willingness to restrict global temperature rise to within 2 °C of pre-industrial levels in collaboration with other nations, such that global greenhouse gas emissions are cut by half by 2050, and rich countries reduce emissions by 80% or more by 2050 COP 15: defining the maximum acceptable increase in global temperature as 2 °C above pre-industrial levels

(continued)

Table 6.1 (continued)

Year	Initiative	Remarks
2010	COP 16	In order to help developing countries implement programs to combat climate change and deforestation, the Cancun Agreements resolved to establish the Green Climate Fund, which would be endowed with $100 billion year starting in 2020
2011	COP 17	Durban Platform for Enhanced Action (ADP) Ad-Hoc Working Group was established.
2012	COP 18	Fundamental pillars of a new global climate accord, which must be signed by 2015 at the latest and will be necessary for all countries starting in 2020.
2013	COP 19	Parties were asked to simulate and communicate intended contributions that were determined at the national level (INDCs)
2014	IPCC Fifth Assessment Report COP 20	The Fifth Assessment Report (AR5) for the IPCC was completed between 2013 and 2014. The Paris Agreement benefited from its scientific contribution COP 20 interpreted the "principle of common again, but differentiated, responsibilities, and respective capabilities" (CBDR&RC) as "CBDR&RC in light of different national circumstances" (CBDR&RC-NC)
2015	COP 21	Establishes the international community's objective of keeping global warming "far below" 2 °C this century. It recognizes the need of rich countries providing financial assistance and technological transfers to poorer countries. The agreement emphasizes the necessity of cities, regions, enterprises, and individuals in achieving this change, in addition to governments
2016	COP 22	Paris Agreement' enforced and authorized a 5-year work plan under which nations would begin to explicitly address the noneconomic effects of climate change. COP 22 also focused on the implementation of the NDCs. The Climate and Development Knowledge Network (CDKN) was also formed, serving as a reference for NDC implementation in LDCs. Many nations urged that the Adaptation Fund should also be applied to the Paris Agreement to guarantee that it remains a political priority
2017	COP 23	Announcing the USA's withdrawal from the "Paris Agreement" Powering Past Coal Alliance was also launched. While the alliance outlines that a coal phase out is needed no later than 2030
2018	COP 24	Provided instructions on how to fulfil Paris Agreement obligations. The rulebook for the Paris Agreement has made progress.

(continued)

Table 6.1 (continued)

Year	Initiative	Remarks
2019	COP 25	Intended to go on with COP 24's implementation of the Paris Agreement's principles and other crucial climate change actions
	European Green Deal	By 2050, the European Green Deal's primary goal is to achieve carbon neutrality. It was accepted by the European Council in December 2019
	IPCC Special Report	Specifically, the Special Report on Climate Change and Land (SRCCL) and the Special Report on the Ocean and Cryosphere in a Changing Climate (SROCC)
2021	COP 26	Committing to end and reverse deforestation, along with cutting methane emissions by 30% by the year 2030
2022	IPCC Sixth Assessment Report	The Sixth Assessment Cycle of the IPCC, which is now in progress, will see the creation of three Special Reports, a Methodology Report, and the Sixth Assessment Report

Source: http://moef.gov.in; https://unfccc.int; https://www.ipcc.ch; https://sustainabledevelopment.un.org; http://www.nihfw.org

the National Council for Environmental Policy and Planning (NCEPP), the Wildlife (Protection) Act of 1972, Environment (Protection) Act of 1986, Forest (Conservation) Act of 1980, the Water (Prevention and Control of Pollution) Act of 1974, the Air (Prevention and Control of Pollution) Act of 1981, the Indian Forests Act of 1927, the Factories Act of 1948, National Forest Policy (NFP), G.B. Pant Institute of Himalayan Environment and Development (GBPIHED); MoEF establishes the Expert Advisory Committee on worldwide ecological concerns, National Wildlife Action Plan, the Biological Diversity Act, National Clean Development Mechanism Authority (CDM), and National Environment Policy (NEP). In 2007, the Prime Minister Council on Climate Change (PMCCC) is established with the following aim:

- Coordinate national action plans for climate change assessment, adaptation, and mitigation
- Advise the government on proactive actions that India might take to address the problem of climate change
- Assist in interministerial cooperation and policy formulation in relevant areas

Direct initiatives to combat climate change were started in 2008 (National Action Plan for Climate Change (NAPCC) 2008). A list of the initiatives taken in India is given in Table 6.2.

Table 6.2 Climate change related initiatives in India

Year	Initiative	Remarks
2004	India's First National Communication to UNFCCC	
2008	National Action Plan on Climate Change (NAPCC)	Details several initiatives that further the nation's development goals as well as its aims for adaptation and mitigation to climate change In order to achieve India's main objectives in the context of climate change, the NAPCC will be implemented through eight National Missions, which will make up the core of the National Action Plan. These missions will contain multifaceted, long-term, and integrated solutions
2009	State Action Plan on Climate Change (SAPCC)	Includes adaptation and mitigation measures to address the effects of climate change, while adaptation has been deemed to be a more crucial component of the plan
2009	Indian Network for Climate Change Assessment (INCCA)	Designed as a scientific network-based program, it is conceptualized to (1) evaluate the causes and effects of climate change using scientific research; (2) once every 2 years, create climate change assessments (greenhouse gas estimations and impact of climate change, associated vulnerabilities and adaptation); (3) create decision-aid systems, and (4) increase capability for managing opportunities and hazards associated with climate change
	National Institute for Climate Change Studies and Actions	Will perform analytical research on climate change-related scientific, environmental, economic development, and technical challenges
2010	India's Greenhouse Gas Emission—2007	Information on India's greenhouse gas emissions for the year 2007 has been updated in this publication
2010	The Expert Group on Low Carbon Strategies for Inclusive Growth was set up by the Planning Commission	To recommend low-carbon routes that are compatible with inclusive growth in India
2010	Climate Change in India: 4x4 Assessment	This assessment combines four significant regions of India: the Himalayan, North Eastern, Western Ghats, and Coastal regions. It examines projected climate and climate change projections for the year 2030 and their effects on four important sectors: agriculture, water, natural ecosystems, biodiversity, and health
2011	The National Carbonaceous Aerosols Programme (NCAP)	Aims to enhance the understanding of the role of carbonaceous aerosols on climate change, to prepare an inventory of black carbon emissions in the country and to assess its impacts on glacier melting
2012	India's Second National Communication to UNFCCC	
2014	Carbon Market Roadmap for India	

(continued)

Table 6.2   (continued)

Year	Initiative	Remarks
2015	Technology Vision 2035	
2015	India's NDC	Describing the climate activities planned under the Paris Agreement
2015	India's First Biennial Update Report to UNFCCC	
2018	India's Second Biennial Update Report to UNFCCC	
2019	First India CEO Forum on Climate Change	
2020	Second India CEO Forum on Climate Change	
2021	India's third Biennial Update Report to UNFCCC	
2021	Rapid Assessment of the CDM and VCM Portfolio—Report of India	
2021	Roadmap for Ethanol Blending in India 2020–2025	

Source: Ministry of Environment, Forest and Climate Change (undated) http://moef.gov.in; https://www.ipcc.ch; http://www.nihfw.org; http://dst.gov.in

6.4.1   National Action Plan for Climate Change (NAPCC)

The NAPCC is the Indian government's management accreditation of climate change and related challenges. It was released by the MoEF in 2008 and is committed to tackling climate change consistently. The NAPCC tackles the country's urgent and vital challenges by altering the development trajectory. The NAPCC categorizes actions that enhance development objectives and compatible co-benefits for successfully tackling climate change (National Action Plan for Climate Change (NAPCC) 2008). The NAPCC and national climate policy were heavily weighted in the 12th 5-Year Plan (2012–2017), which emphasized that climate change concern should infuse all procedures of strategy for any assignment to thrive over the long term; it should have a variety of objectives, enthusiastic execution technology, and adequate financial support (National Research Council 2013; Kumar and Naik 2019). Eight national missions make up the backbone of the NAPCC:

National Solar Mission (NSM)
National Mission for Enhanced Energy Efficiency (NMEEE)
National Mission on Sustainable Habitat (NMSH)
National Water Mission (NWM)
National Mission for Sustaining the Himalayan Ecosystem (NMSHE)

National Mission for Green India (GIM)
National Mission for Sustainable Agriculture (NMSA)
National Mission on Strategic Knowledge for Climate Change (NMSKCC)

6.4.2   National Solar Mission (NSM) and National Mission for Enhanced Energy Efficiency (NMEEE)

The NSM under the NAPCC was designed to support the development and use of solar energy to create power and electricity, with the ultimate goal of competing with fossil-dependent energy supplies. The objective entails re-establishing a solar research center, forming worldwide partnerships in research and technology, improving local manufacturing capabilities, and generating funding and foreign support. The NMEEE seeks to improve energy competence by putting forward cutting-edge commercial ideas in the area of energy efficiency. The NMEEE has started four programs to improve energy efficiency in companies that use a lot of energy: Perform Archive and Trade (PAT), Energy Efficiency Financing Platform (EEFP), Market Transformation for Energy Efficiency (MTEE), and Framework for Energy Efficient Economic Development (FEEED) (www.mnre.gov.in). At COP 21, the GoI reacted quickly to establish India as being completely concerned about global duty regarding climate change problems. To fulfill international duties relating to climate change, the Indian government established a new mission called the "International Solar Alliance" with France. The mission's objective was to improve solar energy reception worldwide, with a focus on the tropics. To fulfill its responsibilities under the NAPCC, the Indian government has also planned to fivefold the nation's capacity to produce solar energy, from 20 to 100 GW by 2022 (Navroz 2020). Renewable energy has been given priority in India's Nationally Determined Contributions (NDCs) and other measures that address the interrelationship between climate change and energy. NDCs place a strong emphasis on decreasing the intensity of GDP emissions by 33–35% from 2005 to 2030 and setting a target of 175 GW of renewable energy by 2022 (National Institution for Transforming India (NITI) 2015).

6.4.3   National Mission on Sustainable Habitat (NMSH)

Plans are being made for a sustainable habitat mission for both current and upcoming climate change mitigation strategies. Additionally, the mission has outlined strategies to handle several aspects of climate issues, including boosting natural resource preservation, energy efficiency, upholding better planning in urban areas, ecological habitation standards, and concurrently tackling climate-related concerns. The Atal Mission on Rejuvenation and Urban Transformation (AMRUT), Swachh Bharat Mission, Smart Cities Mission, and Urban Transport Programme are the four major missions that the GoI ensured were carried out to achieve this goal by the

Ministry of Urban Development. The objective is to manage sustainability while acclimating and reducing greenhouse gas (GHG) emissions. According to the energy and resources institute of the Ministry of Urban Development, if these four NMSH tasks are implemented well, GHG emissions might be reduced by up to 133 million tons of CO_2 equivalent by 2021 and 270 million tons by 2031. (http://cpheeo.gov.in) (CPHEEO undated).

6.4.4   National Water Mission (NWM)

The NWM under the NAPCC focuses on water conservation, appropriate reuse, and motivating people to adopt environmentally friendly water harvesting and protection methods. Through integrated water resource management, the major objective of NWM is to preserve water, reduce waste, and provide a more equal supply both between and within states. The mission's five declared goals are as follows:

Comprehensive water data base in public sphere and assessment of the effects of climatic changes on the water resource
Promotion of individual and state action for water conservation, expansion, and preservation
Consideration of vulnerable areas, including over-exploited areas
Increasing water use efficiency by 20%
Promotion of basin level integrated water resources management.

The key initiatives of the Ministry of Water Resources are the Ganga Rejuvenation, River Interlinking, Command Area Development and Water Management, Flood Management Wing Programs, Research and Development Program in the Water Sector, and Dam Rehabilitation and Improvement Program (www.jalshakti-dowr.gov.in).

6.4.5   National Mission for Green India (GIM)

To offset the threat posed by climate change, the GIM recommends, as part of the NAPCC, protecting and enhancing Indian forest cover. The goal emphasizes various ecological amenities, carbon sequestration, and emissions reduction as co-benefits and anticipates a comprehensive greening potential. To improve ecological community features like carbon sequestration and sustainability (in forests and other biomes), hydrological services, and biological diversity; to provide resources like fuel, food, and forest produce; and to increase the forest-dependent population, the mission's goals are to increase forest cover to 5 million hectares (mha) and improve the quality of forest cover on an additional 5 million hectares (mha) of forest/non-forest lands (www.naeb.nic.in). The strategic plans of four states, Sikkim, Maharashtra, Madhya Pradesh, and Himachal Pradesh, have been approved by the GIM's national executive council. The Ecosystems Services Improvement Project (ESIP), which is being carried out in Madhya Pradesh and Chhattisgarh with support from the World Bank, was introduced by the GIM in 2018. To manage ESIP compliance, the Indian Council of Forestry Research and Education (ICFRE) established a project implementation branch (GoI 2019; www.naeb.nic.in).

6.4.6 National Mission for Sustainable Agriculture (NMSA) and National Mission on Strategic Knowledge for Climate Change (NMSKCC)

The NMSA has been created by combining, integrating, and incorporating all currently running and newly proposed activities/programs connected to sustainable agriculture, with a focus on soil and water conservation, water usage efficiency, managing soil health, and developing rainfed areas. The goal of NMSA will be to promote community-based approaches that judiciously use common resources (www.nmsa.dac.gov.in). The NMSKCC is largely expected to create a thriving and active knowledge system that would support and guide national action for successfully addressing the goal of environmentally sustainable development (Government of India 2010; Aryal et al. 2019).

6.5 Initiatives for Indian Himalayas

The sustainability of the Himalayas is a major concern; hence several policies and action plans for reducing the impact of climate change and preparing for it have been created. To achieve its objectives for ecological preservation and sustainable development in the IHR, the MoEF founded the GBPIHED in 1988. To promote the environmental and economic development of the Himalayas, the organization created an "Action Plan for the Himalayas" in 1992. Many IHR centers are working to create action plans for ecological sustainability and resource management. The Centre for Land and Water Resources Management, the Centre for Socio-Economic Development, the Centre for Biodiversity Conservation and Management, and the Centre for Environmental Assessment and Climate Change are all collaborating to develop environmental concerns for mitigating and adapting strategies to deal with climate change risks in the region (http://gbpihed.gov.in). India continues to advocate for practical, compliant, and acceptable universal ways based on the idea of Common but Differentiated Responsibilities and Respective Capabilities, having made major contributions to the multifaceted UNFCCC talks (CBDR-RC). In order to safeguard the IHR, the GoI has started several initiatives, including the National Mission for Sustaining the Himalayan Ecosystem (NMSHE), Governance for Sustaining the Himalayan Ecosystem (G-SHE), Hill Area Development Programme (HADP), Indian Himalayas Climate Adaptation Programme (IHCAP), the National

Adaptation Fund for Climate Change (NAFCC), and Climate Change Action Program (CCAP), among others.

6.5.1 National Mission for Sustaining the Himalayan Ecosystem (NMSHE)

The NMSHE, a component of the NAPCC, aims to address important issues such as melting glaciers, biodiversity, wildlife, human livelihoods, and planning to preserve the Himalayan ecosystem. The NMSHE's main function is to oversee the provision of food to the Indian Himalayas and to carry out research and studies to create Himalayan region policies. The government enlisted the aid of six task force institutions to accomplish this aim, which included geological richness, water, forest resources, traditional knowledge, and Himalayan agriculture. The NMSHE is making an effort to solve serious concerns including melting glaciers, biodiversity, wildlife, and human livelihoods and plans to preserve the Himalayan ecosystem. The primary objective of the NMSHE is to address governance to protect the Himalayan environment. Creating policies for the Himalayan region to address climate-related challenges and accomplish long-term development objectives is the focus of this study and action effort. These are the main objectives of the mission:

To conserve biodiversity, forest cover, and other ecological values in the Himalayan region
Sustainable development of the country by enhancing the understanding of climate change, its likely impacts, and adaptation actions required for the Himalayas
To facilitate the formulation of appropriate policy measures and time-bound action programs to sustain ecological resilience and ensure the continued provisions of key ecosystem services in the Himalayas
To evolve suitable management and policy measures for sustaining and safeguarding the Himalayan ecosystem along with developing capacities at the national level to assess its health status.

Twelve Himalayan states were strengthened in their ability to create and carry out strategies for addressing climate change, do susceptibility analyses, and increase public knowledge of its consequences by the NMSHE following the cooperative federalism principles. To produce reports on all fronts, the GoI NITI Aayog formed working groups in the IHR mountains in partnership with institutions for sustainable development (National Institution for Transforming India (NITI) 2017). To accomplish sustainable development objectives in the IHR and analyze the reports based on working group recommendations, the NITI Aayog established the "Himalayan State Regional Council" (https://niti.gov.in).

6.5.2   Governance for Sustaining Himalayan Ecosystem (G-SHE)

A comprehensive climate change adaptation strategy must include the G-SHE, which offers suggestions for managing and governing the Himalayan ecosystem. It also covers the main ecological issues raised by the IHR, including water, energy, urbanization, forest management, and tourism (http://moef.gov.in). It aims to provide the residents of the IHR with a solid infrastructure, a good standard of living, and a clean and sustainable environment. To improve both the general quality of life in urban and rural areas, it, therefore, encourages cleanliness. The following programs are part of the diverse G-SHE:

Smart City Mission
Swachh Bharat Mission
Door-to-Door Garbage Collection
Ban on Plastic
Community-Based Ecotourism
Spring-shed Development Programme, etc.

6.5.3   National Mission on Himalayan Studies (NMHS)

The NMHS was formed by the Indian government to research the IHR. To find answers for the long-term growth of the IHR, the NMHS aims to create a knowledge network of experts and institutions. The goal of this mission is to improve people's lives in the IHR following the NEP, 2006 (www.nmhs.org.in). These are the objectives for this operation:

Fostering conservation and sustainable management of natural resources
Enhancing supplementary and alternative livelihoods of IHR peoples and the overall economic well-being of the region
Controlling and preventing pollution in the region
Fostering increased/augmented human and institutional capacities and the knowledge and policy environments in the region
Strengthening, greening, and fostering the development of climate-resilient core infrastructure and basic service assets.

6.5.4   Hill Area Development Programme (HADP)

To address issues in the hill area, the HADP was founded at the beginning of the fifth 5-year plan. The main objectives of this program are to advance environmental development and enhance the socioeconomic standing of those who reside in the hills. As a consequence, by using natural resources lawfully in regions that are protected by the program, the HADP programs have attempted to boost basic life support systems for people in hill areas (https://niti.gov.in).

6.5.5   Indian Himalayas Climate Adaptation Programme (IHCAP)

The Swiss Agency for Development and Cooperation (SDC) launched the IHCAP. The IHCAP sought to advance understanding of climate change issues in the IHR and build the ability of research organizations, planners, experts, and other stakeholders to create and put into action climate change adaptation strategies. As a result, the IHCAP has started taking actions to spread awareness of climate change issues in the IHR and to create strategic organizations to support and assist the state-level implementation of the NMSHE and related action plans in the hilly states. Its objective is to improve the climatology knowledge of both Indian and Himalayan state institutions, with an emphasis on glaciology and related issues as well as policy implementation (http://dest.hp.gov.in/). The program centered on the following objectives:

Strengthening capacities for adaptation planning and implementation in HP through research, training and capacity building

Scientific capacity building in the field of Glaciology and related areas Facilitating dialogues between Himalayan states and key stakeholders for mainstreaming climate change concerns into development planning.

6.5.6   National Adaptation Fund for Climate Change (NAFCC)

The NAFCC was established by the Indian government to support programs and activities for adaptation intended to lessen the effects of climate change on communities and industries. The main objective of the NAFCC is to help disadvantaged states and UTs with the financial burdens of adaptation. The National Bank for Agriculture and Rural Development (NABARD) has been assigned the role of a national implementing agency in charge of carrying out NAFCC adaptation programs. In three blocks of the Sirmaur district of Himachal Pradesh, the MoEFCC, GoI, and NAFCC have approved funding for a climate change adaptation project called Sustainable Livelihoods of Agriculture-Dependent Rural Communities (SLADRC) in the drought-prone district of Himachal Pradesh through climate-smart solutions. The following are the primary goals and projects of NAFCC:

Identification of sectoral adaptation strategies to assist rural communities for implementation

Development of long-term activity-wise action plan

Assessment of community-level vulnerability with exposure, sensitivity, and adaptive capacity different from conventional planning process on Agriculture, Water-irrigation, Crop diversification & livelihood practices

Documentation of best practices being adopted by the farmers
Development of GIS-based information systems to represent impacts of climate change in
    district Sirmaur of HP
Create an enabling framework for climate change adaptive capacity
Training module development on climate-smart approaches
Training/Orientation of target farmers on climate-resilient agriculture/horticulture
Extension services and handholding support to target farmers from time to time
Demonstration of different packages of practices, adaptive to climate variability
Organizing dissemination workshops on project learning.

6.5.7   Climate Change Action Program (CCAP)

The CCAP is a framework developed under the NAFCC in Himachal Pradesh to minimize climate-related vulnerability and improve the adaptive competence of rural small/marginal farmers, particularly rural women, through the establishment of a union of climate-smart agricultural technology in connection with mandated social engineering and capability-building activities, improving food safety and livelihood to raise resilience. The following seven training modules have been created to increase the capacities of rural marginal farmers, particularly women, according to the training demand assessment (http://sladrc.in/).

Drought resilient varieties and cropping systems training
Efficient water management systems including micro-irrigation for water use efficiency
Management of soil nutrition, including practices to enhance soil organic carbon
Efficient pest control through integrated pest management
Agro-met advisory plan as per local weather conditions
Governance aspects of community institutions and convergence
Farmers producers' organization and climate change adaptation.

Bhutan, India, Myanmar, and Nepal are among the nations in the Himalayas that the International Centre for Integrated Mountain Development (ICIMOD) has been collaborating with for many years (https://www.icimod.org). The International Climate Initiative (IKI) is one of the effective tools used to assist climate change mitigation and biodiversity on a worldwide scale. The IKI provides funding for global climate change mitigation and biodiversity conservation in conformity with the UNFCCC and CBD objectives. Over 750 activities related to climate change and biodiversity were approved by the IKI between 2008 and 2020, and over 60 nations received funding from its programs. The efforts of the IKI are intended to put into practice the convention on the rights of children. Initiatives from the IKI are intended to implement the United Nations Sustainable Development (https://www. international-climate-initiative.com).

6.6   India Nationally Determined Contribution (NDC)

India submitted its NDC to the UNFCCC on October 2, 2015, in response to the decisions made by the Conference of the Parties, outlining the climate actions planned following the Paris Agreement. India's NDC has the following eight objectives:

(a) To promote a sustainable, healthy way of life based on the customs and principles of moderation and conservation
(b) To take a road that is more environmentally friendly and clean than the one other has before at the same degree of economic growth
(c) To lower its GDP's emissions intensity by 33–35% from 2005 levels by 2030
(d) To attain a cumulative installed capacity of nonfossil fuel-based energy resources equal to around 40% of total electric power by 2030 with the aid of technology transfer and low-cost international financing, including from the Green Climate Fund (GCF)
(e) To increase the amount of forest and tree cover by 2030, adding 2.5 to 3 billion tons of CO2 equivalent as a new carbon sink
(f) Improving investments in climate change-vulnerable development programs, notably in agriculture, water resources, the Himalayan area, coastal regions, health, and disaster management
(g) To mobilize domestic resources as well as extra funding from developed nations to implement the aforementioned mitigation and adaptation measures in light of the resource gap and resource requirements
(h) To increase resources, establish local and international frameworks, and develop capabilities for cooperative collaborative research and development of such future technologies in India

6.6.1   State Action Plan on Climate Change (SAPCC)

The GoI requested governments to submit plans based on the policy framework of the eight NAPCC missions in response to NAPCC developments to address major climate change problems. Decentralization is necessary since numerous components of these objectives, including agriculture and water, are state issues in India (Dubash and Jogesh 2020). To meet their post-2020 goals, the states and territories have been required to evaluate their SAPCCs and concentrate on their responsibilities (MoEFCC 2019). The Indian Network for Climate Change Assessment (INCCA 2010) set up organizations to look at climatic conditions, develop GHG inventories, and provide a way to coordinate research findings. Additionally, the government has mandated that the NAFCC assist particularly vulnerable states in line with the needs and priorities established by the SAPCC and the pertinent NAPCC missions (Ministry of Environment, Forest and Climate Change 2016). An NAFCC entity called the NABARD offers to fund projects dealing with the environment (Dubash and Ghosh

2020). As a result, the NAPCC and SAPCC, along with the respective state governments, are putting various initiatives into action through missions to solve climate-related problems. The right to life includes a wide variety of rights, according to Article 21 of the Indian Constitution. The right to an environment free from pollution is also protected by this article. The right to a clean environment was made a fundamental right by the Indian legal system, which widened the definition of the right to life. Despite the participation of the government and courts, environmental concerns are spreading since these regulations are not being followed. Future generations, as well as the current generation, are its immediate victims. The environmental laws and regulations of India may need to be changed on a systemic level. The need that these rules to be adequately enforced is the most crucial.

6.6.2 Himachal Pradesh

The Beas River flows in two states of India: Himachal Pradesh and Punjab. Thus, understanding of their policies is very important. The SAPCC goals and their strategies are taken in 2012 and assessed for the climate change action planning period 2021–2030 (Government of Himachal Pradesh 2012, 2021). The state of Himachal Pradesh prepared the SAPCC (2012) with the following missions (Table 6.3):

- Himachal Pradesh Solar Energy Programme
- Himachal Pradesh Energy Efficiency/Saving Programme
- Himachal Pradesh Sustainable Development Programme for Urban and Rural Areas
- Sustainable Water Management
- Sustainable Development to Save the Himalayan Ecosystem
- Programme for Greening of Himachal
- Sustainable Agriculture
- Strategic Knowledge for Climate Change: Towards Carbon Smart Growth

6.6.3 Punjab

The state of Punjab prepared the SAPCC (Jerath et al. 2014) with the following missions (Table 6.4):

- Water Mission
- Sustainable Agriculture Mission
- Green India Mission
- Sustainable Himalayan Mission
- Sustainable Habitats Mission
- Solar Mission
- Mission on Enhanced Energy Efficiency
- Mission on Strategic Knowledge

Table 6.3 SAPCC of Himachal Pradesh with objectives

Mission	Objectives
Himachal Pradesh Solar Energy Programme	Plans are in place in the state of "Himurja" to significantly expand the proportion of solar energy in the state's overall energy mix, while simultaneously noting the need to enhance the availability of other nonfossil options including thermal energy, wind energy, and biomass
Himachal Pradesh Energy Efficiency/ Saving Programme	Launched the "Atal Bijli Bachat Yojna" in the state by providing free CFLs to the people of the state in order to encourage energy conservation and a transition to energy-efficient appliances/equipment Complete prohibition on the use of coal for space heating, etc. Creating economic mechanisms in Himachal Pradesh to encourage energy efficiency Committed to using the whole potential of 22,000 MW of hydropower available in the state, despite the fact that demand in the state is significantly less than the available potential, in order to contribute to the country's clean energy demand in order to reach the established targets for reducing GHG emissions Promote the use of solar passive heating systems as well as biogas facilities Discourage the energy-intensive businesses that contribute significantly to global warming
Himachal Pradesh Sustainable Development Programme for Urban and Rural Areas	Is devoted to developing in a sustainable way in order to preserve its attractive surroundings through advances in solid waste management, wastewater management, mode shift to public transportation, building and road development, energy efficiency in structures, and so on. The State Government is dedicated to encouraging sustainable development and energy efficiency as an integral part of urban and rural planning through a variety of programs
Sustainable Water Management	Clear guidelines are therefore needed in these matters in the State because water-related projects have many socioeconomic aspects and issues, such as environmental sustainability, resettlement and rehabilitation of project-affected people and livestock, public health concerns of water impoundment, dam safety, etc.
Sustainable Development to Save the Himalayan Ecosystem	The State Government will develop management methods to preserve and protect Himalayan glaciers and mountain eco-systems. To assess freshwater resources and ecosystem health in the Himalayan environment, an observational and monitoring network for the Himalayan environment must be built. Cooperation with neighboring countries will be sought in order to expand the network's scope
Programme for Greening of Himachal	Protecting and restoring the forest ecosystem will (1) increase biological diversity, boost water supplies, enable carbon sequestration, meet recreation needs, and support communities that depend on the forest through improved non-wood forest produce and (2) promote successful businesses, luring investors who view sustainability as a workable business model

(continued)

Table 6.3 (continued)

Mission	Objectives
Sustainable Agriculture	To make agriculture more adaptable to climate change, the State Government is rethinking functions in a more efficient manner. It would identify and create novel agricultural types, particularly temperature-resistant crops, and alternate farming patterns capable of withstanding climatic extremes, protracted dry periods, flooding, and changing moisture availability
Strategic Knowledge for Climate Change: Towards Carbon Smart Growth	A 3-year "Community Led Assessment, Awareness, Advocacy, and Action Programme (CLAP) for Environment Protection and Carbon Neutrality in H.P." has been launched in the state with the goal of developing Himachal Pradesh as a sustainable and climate-resilient state by mobilizing community responsibility for environmental assessment, environmental protection, and carbon neutrality. The initiative would ensure high-quality knowledge and a targeted approach to many facets of climate change, including the socioeconomic implications of climate change on health, demography, migratory patterns, and rural community livelihoods
Watershed Development Programmes (WDP)	To promote soil moisture conservation, rainwater collection, afforestation, pasture development, sustainable agriculture and horticulture practices, and land development to boost the productive potential of watersheds and their related natural resource base. Additionally, it establishes and strengthens community-based institutional frameworks for long-term resource management, offers non-farm sector training and employment opportunities and ensures village communities' involvement in. participatory planning, implementation, and social and environmental management
Himachal Pradesh Crop Diversification Project	To assure organic farming and vegetable production, as well as to build infrastructure for agricultural growth. It is carried out in partnership with the Japan International Cooperation Agency (JICA)
Organic Farming Policy	It takes into account every aspect of organic farming, including the vision, mission, strategy for desired policy revision, awareness-raising among stakeholders, support for farmers through organic technology and extension, quality assurance of state produce, organic inputs, demand and supply issues, developing organic supply chains, governance, and implementation
Horticulture Technology Mission	With the goal of integrating horticultural growth in Himachal Pradesh.
Pandit Deen Dayal Kisan Bagwan Samridhi Yojna	An all-encompassing plan being undertaken by the Government of Himachal Pradesh to provide self-employment possibilities and diversify farming in order to boost farmers' economic situation

(continued)

Table 6.3 (continued)

Mission	Objectives
Organic Policy	Recognizing and encouraging the organic industry in the state Creating an enabling environment in the state for organic farming by adopting suitable regulations, strategies, and support services for organic agriculture Create favorable policies and programs to transform Himachal into an organic compost-rich state Take measures to recognize woods, grazing grounds, and pastures as organic, certified/uncertified places Foster an investment climate for organic agriculture and organic agrotourism
Bio-carbon, Clean Development Mechanism (CDM) Project	Will go a long way toward protecting and supporting our commitment to environmental preservation and protection by sequestering greenhouse gases (GHG) by developing forestry plantings on primarily degraded land
Integrated Catchment Area Treatment Plan	To guarantee a scientific and need-based approach to catchment treatment in which all stakeholders may participate in the development of catchment areas Cumulative mitigating strategies for landslide and soil erosion concerns To address the issue of silt and debris load in the long run Monitoring the sediment load from streams that discharge directly into the reservoir Combining scouring/sloughing and slide prevention for straight draining catchments

6.7 Summary

The environmental problem of climate change in the basin will have an impact on both the upstream and downstream populations' quality of life. The UNFCCC and the KP are the legal pillars of the existing international climate governance system. As a result, pragmatism seems to define the current situation of Indian climate change policies. India's current climate change policy has undergone a significant transition, and it now centers on bold initiatives on both the national and international levels. To accomplish "multipronged, long-term integrated solutions" in the context of climate change, the NAPCC created eight national missions. The MoEFCC collaborates with many ministries involved in electricity, renewable energy, urban development, research and technology, water resources, and agriculture to combat these growing climate-related challenges. The GoI is creating and carrying out many strategies to increase the country's capacity for producing solar energy. Other missions organized by the Indian government include Atal, Swachh Bharat, and Smart Cities for Sustainable Habitat. Reports on the successful execution of the missions were written by the Energy and Resources Institute (TERI). Countrywide reforestation efforts are being made by the GIM. To prepare for and mitigate climate change in the Himalayan region, the NMSHE engaged all 12 of states. To guarantee the IHR's long-term survival, many organizations—including the NMHS, IHCAP, and NMSHE G-SHE—have a lower GHG emission target in

Table 6.4 SAPCC of Punjab with objectives

Mission	Objective
Water Mission	By tackling the effects of climate change on water resources, the state intends to take an integrated strategy to conservation and management of its water resources, increase water usage efficiency, reduce water pollution, limit waste, and ensure fair distribution of water across the state
Sustainable Agriculture Mission	Bring about the second green revolution through sustainable agriculture practices, and thereby assure food security in a changing climatic situation. Crop diversity, optimal resource usage, judicious use of technology, and contributions from fresh research will all be part of the strategy. Encourage the use of agricultural waste to generate electricity
Green India Mission	Boost the state's green cover to 15% of its land area by 2022, improve planting in degraded forests, and increase the income generated by ecosystem services supplied by forests
Sustainable Himalayan Mission	Conserve flora, animals, wetlands, agriculture, and forest biodiversity to ensure the sustainability of the Shivalik ecosystem
Sustainable Habitats Mission	The state intends to create regulations and procedures that will allow habitats to adapt to climate change problems Identify and execute measures in metropolitan areas to mitigate the heightened heat island effect and to manage municipal solid waste and transportation in a sustainable manner
Solar Mission	By 2022, the solar energy mix must be increased by at least 2000 MW
Mission on Enhanced Energy Efficiency	To enhance energy efficiency by 3–7% in big energy consumers recognized by BEE and 15–20% in the SME sector
Mission on Strategic Knowledge	The Punjab mission on strategic knowledge intends to increase awareness of climate change processes, their consequences for various sectors, and vulnerabilities associated with them in order to sustainably adapt to climate change and reduce climate change causes
Punjab State Water Policy	Envisions that available water resources should be used effectively and wisely to satisfy drinking water and agricultural demands while also encouraging conservation and community engagement
Pollution Control	The Punjab Pollution Control Board established in 1975 enforces the national environmental laws and acts such as the Water (Prevention and Control of Pollution) Act, 1974, Amended 1988; Water (Prevention and Control of Pollution) Cess Act, 1977; the Environment Protection Act, 1986; and the associated rules; Water (Prevention and Control of Pollution) Cess Rules, 1978
Agriculture Policy	Increase in the rate of seed replacement Prudent and balanced fertilizer application based on soil testing Nutrient and pest control that is integrated Increased production and promotion of the cattle sector Water resource conservation, development, and management that is long term Promotion of effective on-farm water management techniques Improving Strategic Agricultural and Livestock Research Efficient transmission of cutting-edge technologies to farmers Reduce difficulties caused by agricultural residue burning

(continued)

Table 6.4 (continued)

Mission	Objective
Punjab Forest Policy, 2008	Sustainable forestry methods are based on fundamental principles of forest management and the application of current technology and scientific knowledge Development of an appropriate Protected Area Network for the protection, conservation, and improvement of the state's wildlife and remaining biological resources Promoting non-timber forest uses such as ecotourism, non-timber forest products, medicinal plants, and biodiversity Practicing socially inclusive forestry and enlisting the collaboration and involvement of rural and forest-dependent communities, as well as other stakeholders, in the process of greening the state People will be given technical help, financial incentives, and extension services to promote social forestry, agroforestry, and tree farming for land use diversification Implementing government policies and programs by using creative techniques to maximize the state's and its citizens' social, economic, and environmental advantages from the forestry industry
Punjab Ecotourism Policy, 2009	(a) Identification and promotion of ecotourism prospective sites; (b) development of permissible and permissible ecotourism sites; and (c) development of permissible and permissible ecotourism sites, ecotourism infrastructure that is environmentally sound; and (c) diversification of tourist activities available at sites; (d) development and enforcement of ecotourism rules and principles activities; (e) ensuring the participation of local populations living in and reliant on the periphery as well as other fields for a living; (f) raising awareness among the general public, local communities, and government employees; (g) outlining a mechanism for securing the partnership of private sector enterprises committed to the goals of ecotourism for the development of infrastructure and services; and (h) sensitization of the community and augmentation of local livelihood through the ecotourism route
National Urban Transport Policy 2006	Reduce travel demand by improving the integration of land use and transportation planning. Allow for more equitable road space allocation; reduce traffic congestion; improve public transportation and road safety; implement integrated transportation systems; and make provisions for the usage of non-motorized vehicles. Encourage the deployment of sustainable technologies to minimize reliance on fossil fuels and Capacity building at the individual and institutional levels, as well as raising awareness for a seamless transition to cleaner technologies, will be undertaken, as will the development of novel funding structures with more participation from the private sector
Punjab New and Renewable Sources of Energy (NRSE) Policy, 2012	To optimize and increase the percentage of innovative and renewable energy sources in the state's total installed power capacity to 10% by 2022 To promote renewable energy programs for satisfying rural energy/lighting demands and augmenting energy needs in urban, industrial, and commercial sectors

(continued)

Table 6.4 (continued)

Mission	Objective
Punjab Energy Conservation Action Plan	To proactively carry out all of the requirements of the EC Act in collaboration with BEE, the state government, and other stakeholders
	Advance the cause of energy efficiency by targeting all commercial energy sources (e.g., coal, liquid petroleum gas, oil, and electricity)
	Reduce energy consumption in generation, transmission, and distribution by implementing end-user DSM (demand-side management) programs and large-scale end-user energy efficiency improvements, as well as logical and prudent usage
	Address utilities' challenges, including as demand and power shortages, through targeted DSM programs
	Encourage the reduction of GHG emissions in the province of Punjab
	Encourage the adoption of energy-saving technology, equipment, processes, and appliances
	Raise knowledge of the EC Act, energy efficiency, standards, best practices, and so forth
	Raise awareness in the state about national energy efficiency initiatives such as the Energy Conservation Building Code, which promotes energy efficiency in buildings, and standards and labelling, which promote the manufacturing and use of energy-efficient appliances
	Reduce domestic energy usage through user education and awareness building

mind. India has significant obstacles to responding to climate change in an effective manner, despite several attempts. Through SAPCC, Punjab and Himachal Pradesh are also contributing to the improvement of the basin's circumstances. To adapt to and mitigate climate change, it is imperative to develop highly effective strategies and monitoring of their implementation.

References

Adams B, Schuurman FJ (1993) Sustainable development and the greening of development theory. Zed Books, London, pp 207–220

Aryal JP, Sapkota TB, Khurana R, Khatri-Chhetri A, Rahut DB, Jat ML (2019) Climate change and agriculture in South Asia: adaptation options in smallholder production systems. Environ Dev Sustain 22:5045–5075

Atteridge A, Shrivastava MK, Pahuja N, Upadhyay H (2012) Climate policy in India: what shapes international, national and state policy? Ambio 41:68–77

Bernauer T, Schaffer L (2010) Climate change governance. ETH Zurich Centre for Comparative and International Studies, University of Zurich, Zurich, pp 1–29

Brisman A (2011) Stockholm conference, 1972. In: Chatterjee DK (ed) Encyclopedia of global justice. Springer, Dordrecht. https://doi.org/10.1007/978-1-4020-9160-5_655

Carter A (2001) Presumptive benefits and political obligation. J Appl Philos:229–243

Drexhage J (2008) Climate change and global governance. International Institute for Sustainable Development (IISD)

Dubash NK, Ghosh S (2020) National climate policies and institutions. In: Navroz DK (ed) India in a warming world: integrating climate change and development. Oxford University Press, New Delhi, pp 329–348

Dubash NK, Jogesh A (2020) State climate change planning in India from margins to mainstream. In: Navroz DK (ed) India in a warming world: integrating climate change and development. Oxford University Press, New Delhi, pp 349–369

Edenhofer O, Wallacher J, Lotze-Campen H, Reder M, Knopf B, Muller J (2012) Climate change, justice and sustainability: linking climate and development. Springer, Dordrecht

Government of Himachal Pradesh (2012) State action plan on climate change 2021–2030. Version II. Department of Environment, Science and Technology. Available via http://dest.hp.gov.in/sites/default/files/PDF/SAPCC%20Web.pdf

Government of Himachal Pradesh (2021) State action plan on climate change 2021–2030. Version II. Department of Environment, Science and Technology. Available via http://dest.hp.gov.in/sites/default/files/PDF/SAPCC%202021-30-web.pdf

Government of India (2010) National mission on strategic knowledge for climate change under national action plan on climate change. Mission document. Department of Science & Technology, Ministry of Science & Technology, New Delhi. Available via https://dst.gov.in/sites/default/files/NMSKCC_mission%20document%201.pdf

INCCA (2010) Climate change and India: a 4x4 assessment. Ministry of Forest and Environment, Government of India

Intergovernmental Panel on Climate Change (IPCC) (undated) Reports of 1990; 1995; 2001; 2007; 2014. Available via https://www.ipcc.ch/

IPCC (2014) Climate change 2014: Synthesis report. Contribution of Working Groups I, II and III to the Fifth Assessment Report of the Intergovernmental Panel on Climate Change [Core Writing Team, Pachauri RK, Meyer LA (ed)]. IPCC, Geneva

Jerath N, Ladhar SS, Kaur S, Sharma V, Saile P, Tripathi P, Bhattacharya S, Parwana HK (2014) Punjab state action plan on climate change. Punjab State Council for Science and Technology and GIZ (Deutsche Gesellschaft for Internationale Zusarnmenarbeit GmbH – German International Cooperation, India), p 329. Available via https://moef.gov.in/wp-content/uploads/2017/09/Punjab.pdf

Knieling J, Filho WL (2013) Climate change governance: the challenge for politics and public administration, enterprises and civil society. In: Knieling J, Filho WL (eds) Climate change governance. Climate change management. Springer, Berlin/Heidelberg, pp 1–5

Kumar P, Naik A (2019) India's domestic climate policy is fragmented and lacks clarity. Econ Polit Wkly 54(7):1–3

Lolaksha NP, Anand A (2017) Climate change and its impact on India: a comment. NLUO Law J 4:81–97

Meadowcroft J (2009) Climate change governance, World Bank policy research working paper (4941)

Messner PD (2017) Game theory in international climate change negotiations: why does the US plan to withdraw from the Paris Agreement and what are the consequences for countries and future climate protection? Doctoral dissertation

Ministry of Environment, Forest and Climate Change (2016) Annual report 2015–2016. Government of India

Ministry of Environment, Forest and Climate Change (undated) Annual Report 2008–2009; 2012–2013. Government of India. Available via http://moef.gov.in.

National Action Plan for Climate Change (NAPCC) (2008) Ministry of Environment, Forest and Climate Change, Government of India. Available via http://pmindia.nic.in/Pg01-52.pdf

National Institution for Transforming India (NITI) (2015) Report of the expert group on 175GW RE by 2022. Government of India. Available via https://www.niti.gov.in/sites/default/files/energy/175-GW-Renewable-Energy.pdf

National Institution for Transforming India (NITI) (2017) Draft National Energy Policy. Government of India, New Delhi. Available via https://www.niti.gov.in/writereaddata/files/document_publication/NEP-ID_27.06.2017.pdf

National Institution for Transforming India (NITI) (2020) Sustainable development in the Indian Himalayan region. Government of India. Available via https://niti.gov.in/sustainable-development-indian-himalayan-region

National Mission on Sustainable Habitat Central Public Health & Environmental Engineering Organisation (CPHEEO) (undated) Ministry of Housing and Urban Affairs, Government of India. http://cpheeo.gov.in/cms/national-mission-on-sustainable-habitat.php

National Research Council (2013) Climate and social stress: implications for security analysis. National Academies Press, Washington, DC

Navroz DK (2020) India in a warming world: integrating climate change and development. Oxford University Press, New Delhi

Ostrom E (2010) A multi-scale approach to coping with climate change and other collective action problems. Solutions 1(2):27–36

Renate M (2004) Governance theory as an advanced control theory? Max Planck Institute for the Study of Societies, Cologne

Sahu AK (2019) The democratic securitization of climate change in India. Asian Polit Policy 11(3):438–460

Saryal R (2018) Climate change policy of India: modifying the environment. South Asia Res 38:1–19

Stern N, Stern NH (2017) The economics of climate change: the Stern review. Cambridge University Press, Cambridge

United Nations (1992) The United Nations framework convention on climate change. Available via https://unfccc.int/resource/docs/convkp/conveng.pdf

Wilde JH (2008) Environmental security deconstructed. In: Brauch HG, Spring UO, Mesjasz HC, Grin J, Dunay P, Behera NC, Chourou B, Kameri-Mbote P, Liotta PH (eds) Globalization and environmental challenges, Hexagon series on human and environmental security and peace. Springer, Berlin/Heidelberg, pp 595–602

Chapter 7
Summary and Conclusion

Abstract The Beas basin has warmed during the last four decades. This is consistent with rising global air temperatures, even if the magnitude of such trends is not very large. Yet over time, even little variations in temperature in a place like the Himalayas might affect the availability of water sources. Seasonal influences on streamflow were expected to outweigh yearly effects owing to variations in temperature and land cover. There were observable variations in streamflow during the winter, pre-monsoon, and monsoon by the late twenty-first century (2071). Future winter, pre-monsoon, and monsoon seasons might benefit from an increase in mean flow for the population. The National Action Plan for Climate Change (NAPCC) is one of several programs that have been implemented at the international, national, and local levels to address the effects of climate and land use/land cover changes. Nonetheless, community involvement is necessary for the successful execution of all the measures.

Keywords Beas River · Climate change · Land use/land cover · Discharge · Policy · Mitigation · Community

The presence of water resources affects a region's economy either directly or indirectly. The function of water resources in the mountain ecosystem is significant for both the upstream and downstream populations' ability to sustain their way of life. Changes in the climate and land cover are thought to be major determinants of how catchments behave hydrologically because they may affect the components of the hydrological cycle and the availability of water supplies. Evidence of the effects of climate variability and land cover changes on hydrological parameters can be seen at both the local and global levels. Additionally, the issue is getting worse because of increased water consumption, which is also rising as a result of the growing population. The Beas basin, where surface water is the primary supply of water, also shows signs of warming. Studies revealed a decrease in river discharge at several locations around the basin, while the degree of the decline varied. A review of the

© The Author(s), under exclusive license to Springer Nature
Switzerland AG 2023
S. Rani, *Climate, Land-Use Change and Hydrology of the Beas River Basin, Western Himalayas*, Advances in Asian Human-Environmental Research,
https://doi.org/10.1007/978-3-031-29525-6_7

literature found the absence of a thorough investigation that considers the relationship among climatic factors, land cover, and streamflow. To effectively plan for and manage water resources, it is necessary to understand how changes in the climate and land cover will affect both current and future water resources. This can be done by using a reliable hydrological model. As a result, the current study is being conducted in the Beas River basin to examine the variation in climatological factors from 1980 to 2020. Understanding the changes in land cover in the basin was another goal of the study. Attempts were also made to examine the impact of changing climate and land cover on the basin's flow regime by the late twenty-first century. The following is a summary of the study's findings:

7.1 Climate Variability Assessment

7.1.1 Air Temperature

The Beas basin is located in the Himalayas' high-elevated topography and is a source of livelihood, highlighting the importance of assessing changes in climate variables. Between 1980 and 2020, the annual T_{mean} in the Beas basin was 14.93 °C. During the study period, 2016 was the warmest year, with an annual T_{mean} of 16.26 °C, and 1997 was the coolest, with an annual T_{mean} of about 13.84 °C. The standard deviation and coefficient of variation of the annual T_{mean} were 0.57 °C and 3.79%, respectively, indicating very minor variation in the annual T_{mean}. During the study period, the annual T_{mean} shows a significant warming trend at a rate of 0.27 °C/decade.

Season-wise, the observed T_{mean} in the Beas basin was 7.15 °C in the winter, 20.12 °C in the pre-monsoon, 21.73 °C in the monsoon, and 12.76 °C in the post-monsoon seasons, respectively, between 1980 and 2020. The winter season T_{mean} showed the highest variation (12.34%), followed by the post-monsoon (6.02%), the pre-monsoon (4.9%), and the monsoon season (2.31%). T_{mean} demonstrated a strong warming trend in the study area during the monsoon (0.25 °C/decade) and post-monsoon (0.35 °C/decade).

Between 1980 and 2020, the basin's monthly T_{mean} varied from 4.24 °C in January to 22.90 °C in June. From June through September, the coefficient of variance reveals a comparatively little amount of variation. The analysis of monthly trends reveals a statistically significant upward trend in T_{mean} for April and August to September. T_{mean}'s magnitude of change varies from 0.17 °C/decade (August) to 0.56 °C/decade (April). The rising tendency in the months' mean T_{min} and T_{max} values may be the cause of the basin's current warming trend.

7.1.2   ET

Between 1980 and 2020, the Beas basin experienced an annual ET of 28.68 mm. Throughout the study period, there is no discernible pattern in the yearly ET. For the Beas basin, the observed ET throughout the winter, pre-monsoon, monsoon, and post-monsoon seasons was 6, 9.05, 9.73, and 2.44 mm, between 1980 and 2020 respectively. Pre-monsoon ET in the studied area exhibited a substantial falling tendency (−0.04 mm/decade) for the study. Additionally, from 1980 to 2020, the basin's monthly ET varied from 4.24 mm in January to 22.9 mm in June. During the study period, the seasonal trend analysis of ET reveals a statistically significant declining trend in April (−0.07 mm/decade) and September (−0.05 mm/decade) and a statistically significant rising trend in December (0.03 mm/decade).

7.1.3   Cloud Cover

In the Beas basin, the annual low cloud cover averaged 14.90% from 1980 to 2020, while the yearly averages for the middle, high, and total cloud cover were 20.07%, 26.6%, and 0.27% respectively. No significant trend in cloud cover over the study period is shown by the annual trend analysis.

From 1980 to 2020, the winter months in the Beas basin had the highest low cloud cover, followed by the monsoon. Winter and monsoon seasons both had the highest and nearly equal values of middle cloud cover. Winter has the largest concentration of high cloud cover, followed by pre-monsoon. High cloud cover in the study area showed a considerable declining trend in the winter (−1%/decade) among all cloud categories. However, there was a noticeable 5%/decade increase in overall cloud cover throughout the winter.

Between 1980 and 2020, the monthly low cloud of the basin varied between 4.90% in November and 36.44% in August. In the study area, middle cloud cover values range from 30.66% in July to 7.51% in October. High cloud cover in the basin also indicates a similar condition. Low cloud cover's monthly trend analysis reveals a statistically significant rising trend in January and a declining trend in November. High cloud cover, on the other hand, exhibits a rising trend in Nov during the study period. During the study period, the total amount of clouds exhibits an increasing tendency in March, a decreasing trend in November, and no trend in December.

7.1.4   Cloud Base Height

Between 1980 and 2020, the Beas basin experienced an average CBH of 2612 m (above the surface). Throughout the study period, there is no significant trend in the annual CBH. Pre-monsoon has the highest CBH in the Beas basin (1980–2020), followed by winter, post-monsoon, and monsoon. The findings revealed an insignificant increasing trend in the study area's seasonal CBH during the study period. In April, CBH has increased at a statistically significant rate of 104 m/decade during the study period.

7.1.5   Precipitable Water Vapor

The PWV in the Beas basin averaged 100 mm between 1980 and 2020. During the study period, the annual PWV exhibits a noticeable rising trend at a rate of 1.56 mm/ decade. Between 1980 and 2020, the observed PWV in the Beas basin during the winter, pre-monsoon, monsoon, and post-monsoon seasons was 20.2, 23.3, 43.3, and 10.6 mm, respectively. During the study period, PWV revealed a significantly rising trend in the study area during the winter (0.42 mm/decade), monsoon (0.54 mm/decade), and post-monsoon (0.22 mm/decade). In the basin, the monthly PWV varied from 4.55 mm in January to 15.12 mm in July between 1980 and 2020. The PWV reveals a statistically significant rising trend in September (0.32 mm/ decade), October (0.39 mm/decade), and November (0.18 mm/decade).

7.1.6   Amount of Rainfall

In the basin, the mean annual rainfall was 46.77 mm from 1980 to 2020. The annual trend analysis for the region reveals no change in annual rainfall from 1980 to 2020. In addition to analyzing annual rainfall, seasonal rainfall trend analysis is carried out to comprehend the seasonal variations in the basin's rainfall. Between 1980 and 2020, the observed mean rainfall during the winter, pre-monsoon, monsoon, and post-monsoon seasons was 12.4, 7.8, 25, and 2.3 mm respectively. There is no significant trend in rainfall according to the seasonal trend analysis. In the basin, the mean monthly rainfall varied from 0.8 mm in November to 11.4 mm in July between 1980 and 2020. It demonstrates that the monsoon season's month of July saw the highest rainfall. The analysis of monthly trends reveals that the rainfall in March, April, May, and July is significantly decreasing. March has a comparatively large amount of change in rainfall. If these magnitudes of a falling trend continue in this manner, it could have serious long-term effects on the Beas River's hydrology.

To comprehend changes in the percentage share of seasonal variability in the rainfall in the basin, the trend in seasonal precipitation is evaluated. In the period between 1980 and 2020, the observed mean rainfall percentage share throughout the winter, pre-monsoon, monsoon, and post-monsoon seasons was 3%, 26%, 53%, and 5%, respectively. This shows that the monsoon season had the largest share. From 1980 to 2020, the average monthly rainfall varied from 1.8% in November to 24.6% in July. It demonstrates the month of July has the largest contribution to annual precipitation. The analysis of monthly trends reveals a notable increasing tendency in the rainfall in August. The findings indicate that, despite large changes in rainfall amounts over the study period, there has been no significant change in the percentage share of rainfall on the monthly scale. The level of relationship of T_{mean} with other variables in the study area is shown by the correlation coefficients (r) result. When compared to T_{mean}, ET exhibits a large positive link between February and July but a negative relationship in April and May. In February through May, July, and December, low cloud cover exhibits a negative association with T_{mean}, while it also exhibits a negative relationship in September. CBH and T_{mean} have a negative relationship in September, but a positive relationship in February through April and December. In March through May, July, and November, rainfall has a negative relationship with T_{mean}.

7.2   Changes in Land Use/Land Cover

According to the spatiotemporal analysis of LULC, open forest and agriculture made up the majority of the study area in both years, followed by dense forest. Less than 2% and 3% of the basin's total area were covered by glaciers/snow and water bodies, respectively, in 2001 and 2020. The upper Beas basin is predominately dense woodland, barren, and natural herbaceous up to Pandoh Dam. The lower Beas basin is largely made up of agriculture, while the middle areas of the Beas basin is dominated by open forest. The lower basin, including the Punjab plains, which is also a state experiencing a green revolution, is seeing a faster spread of agricultural land. It shows that among surface features, forests, deserts, and agricultural land have the greatest influence on surface flow.

When compared to the total area, the built-up area expanded from around 0.38% to 1.28% in 2014. In 1990 and 2000, respectively, the built-up area was 0.84% and 0.93% of the total basin area. The lower Beas basin in the Punjab plains saw the greatest growth in settlement density. Rural settlements make up the majority of the basin's spatial distribution of settlement density (97% in 1975 and 93% in 2015), whereas urban areas make up less than 5% (1.3% in 1975 and 3.5% in 2015). Between 1975 and 2015, there was a dramatic increase in the middle and lower parts of the Beas basin's population density.

During the study period, the LULC classes in the upper Beas basin has changed. Between 1980 and 2020, there were two major trends in LULC: an increase in cultivated land (344 km²) and built-up area (70 km²) and a forest decline (268 km²). The water body under observation had a little change of roughly 3 km². Snowfall decreased during the time frame as well; however, it is challenging to draw any conclusions about a trend from images taken across two different periods. Concern exists regarding the ongoing destruction of forests between 1980 and 2020. From 1980 to 2020, the built-up area would be converted to agricultural land, forest, grassland, BUW, and water bodies at a rate of 36.3%, 17.2%, 11.3%, 27.8%, and 5.3%, respectively. A total of 35.6% of cultivated land has been turned into grassland, 22.5% into the desert, and 0.3% into water bodies. A total of 1.2% of forestland has been converted to agriculture, 9.3% to grassland, 11% to BUW, and 0.8% to water bodies. About 4.4% of the area of grassland has been turned into arable land, 13.3% into woodland, and 20.6% into cultivated land. A little over 3% of water bodies have been turned into built-up areas, 13.3% into cultivated land, and 16.7% into barren. 13.6% of the snowpack has turned barren.

According to the snow cover area (SCA) assessment, the mean monthly SCA for the years 2001–2020 shows that snow accumulates from late October to mid-March before declining from April until the end of the year. SCA peaked in February and peaked in September, respectively. Compared to other land cover groups, SCA is a more dynamic land cover. The Beas basin's mean monthly change in SCA shows snow accumulation and ablation in various elevation zones between 2001 and 2020. Most of the land above 3000-m AMSL is covered by SCA. Based on elevation, the study area's SCA distribution for each month was evaluated. Snow coverage reaches a height of 2000 m in January and stays at or above 4000 m in August. The trend in the study area's SCA does not indicate a significant trend in the study area. However, it shows a decreasing trend in the pre-monsoon during the study period.

7.3   Effect of Climate and Land Cover Changes on Flow

Two important elements that affect a watershed's hydrological behavior are the climate and LULC. The SWAT hydrological model, which accounts for several elements like climate, land cover, soil, slope, and elevation, was used in the current study to estimate the availability of water resources. The model was parameterized to regionalize it and make it suitable for the basin. Using the SUFI2 method in the SWAT-CUP program, a total of 18 parameters relating to elevation band, snow, and hydrology were chosen for the model calibration. Overall, the basin's simulated mean monthly flow performance from the model is satisfactory. Eleven potential scenarios for climate and land cover change were examined to determine how often streamflow in the upper Beas basin might fluctuate in the near future (2071). The study has selected eight climate change (CC1–CC8) scenarios after looking at the analyses of air temperature, rainfall, and predicted changes in the climatic variables. Three scenarios (LC1–LC3) were chosen based on SCA descriptive data to anticipate how the flow will vary as SCA changes.

The baseline period of 1974–2020 was used to examine the basin's general hydrological conditions. At Thalout, the monthly, seasonal, and annual mean flows were estimated to be significantly declining. The overall correlation results showed that the study period's lowering flow at the station was influenced by the rising mean air temperature. However, there is a need to investigate additional basin-level variables that may have an impact on the flow during the baseline period.

According to air temperature scenarios (CC1–CC2), the winter season has the largest increase in flow, which is followed by pre-monsoon by 2071. A change in precipitation from snowfall to rainfall or an increase in snowmelt under warmer climate conditions could increase winter flow. Furthermore, these two processes could boost streamflow throughout the basin's winter season at the same time. The spike in snowmelt runoff due to the warmer environment may be the cause of the expected increase in flow during the pre-monsoon season. The populace would benefit from higher anticipated flows during this season because there is a greater need for hydropower to power agriculture at this time of year. In general, a change in peak streamflow caused by the spring melting condition would have an impact on the time of the predicted peak streamflow by 2071.

In addition, the model has predicted a significant increase in flow during the pre-monsoon and post-monsoon relative to baseline under climate change scenarios (CC3 and CC4) when a change in rainfall is taken into account. The early melting of the snow could be a cause of increase in pre-monsoon discharge of the basin.

A significant increase in flow is anticipated during winter, followed by pre-monsoon, in climate change scenarios (CC5–CC8) where both air temperature and rainfall change simultaneously. It is due to the wintertime switch from snowfall to rainfall in the form of precipitation. Snowmelting rise during these seasons due to the warmer temperature. Under all climate change scenarios, changes in the basin's winter-predicted mean streamflow would range between 0.28% and 1.23%, while changes in the basin's monsoon-predicted mean flow would range between −0.11% and 0.45% in the near future (2071) relative to the baseline. It could be related to an increase in mean air temperature, which would cause early snowmelt and a switch from snowfall to rainfall as a result. Under all climate change scenarios, the predicted mean annual streamflow would vary from baseline by 0.08–0.66% by 2071.

By the end of the century, monsoon streamflow of the basin would have changed by a range of −13% to −2.80% under LULC scenarios. Although due to variations in the amount of snow cover in the basin, the decrease in streamflow during pre-monsoon in the basin would range from −5.37% to −23.78% by 2071. In the basin, the percentage change in mean annual flow would range from −5.88% to −2.41%. Because rain and snowmelt both contribute to the flow during the monsoon, snowmelt is the principal driver of the flow before the monsoon. The amount of snowmelt in the basin would certainly decrease in the near future if snow cover were to decrease. For the baseline and forecast periods in each of these three scenarios, the time of the expected peak streamflow remains unaffected in the hydrograph; however, the magnitude of the peak streamflow is reduced due to spring melting conditions.

Overall, the results show that the Beas basin has been warming over the past four decades. This is in agreement with increases in global air temperature; albeit the amplitude of those trends is not huge. Over time, however, even small temperature changes in a region like the Himalayas could have an impact on the availability of water supplies. It was anticipated that seasonal effects would outweigh annual effects on streamflow due to changes in temperature and land cover. By the late twenty-first century (2071), there were noticeable fluctuations in streamflow during the winter, pre-monsoon, and monsoon. The population would benefit from an increase in mean flow during the winter, pre-monsoon, and monsoon seasons in the future. Several initiatives have been taken at international, national, and local levels to combat the impact of climate and LULC changes including the National Action

Plan for Climate Change (NAPCC). However, there is a need for community participation for the effective implementation of all the programs.

The findings of the current study might have been improved by the availability of more substantial quality data of climate, snow/glaciers, and river discharge. The current study has made three contributions to the body of knowledge. To effectively manage water resources in the Beas basin, researchers and planners can use the study's precise basic data, such as elevation, climate, soil, and land cover and projected discharge in the hydrological model. Furthermore, the current work has developed decision tree classification rules for the different Landsat images (MSS, TM, and OLI TIRS) that will assist researchers in producing LULC maps of a particular area. Last but not least, relatively few studies have looked at how variations in climate and land cover impact streamflow in the upper Beas basin. The objective of the current study was to fill this knowledge gap by using the SWAT model. It included a comprehensive analysis of the hydrology of the upper Beas basin's sensitivity to various climate and land cover change scenarios by the late twenty-first century. Finally, I'd want to use a Tim Flannery quotation to wrap up this book (Flannery 2006):

> What we need now is good information and careful thinking, because in the years to come this issue will dwarf all the others combined. It will become the only issue. We need to re-examine it in a truly skeptical spirit- to see how big it is and how fast it's moving. So that we can prioritise our efforts and resources in ways that matter.

Reference

Flannery TF (2006) The weather makers: how man is changing the climate and what it means for life on earth. Grove Press, New York